200 YEARS AND GROWING:

THE STORY OF STARK BRO'S NURSERIES & ORCHARDS CO.

LEANN ZOTTA

Stark Bro's Nurseries & Orchards Co.
Attn: Book Permissions
PO BOX 1800
Louisiana, Missouri 63353

To inquire about ordering bulk hardcover copies, please contact:
Stark Bro's Nurseries & Orchards Co.
Attn: Book Sales
PO BOX 1800
Louisiana, Missouri 63353

First edition published 2015.
BOOK DESIGN BY LEANN ZOTTA
EDITED BY KEN LANE

With significant contributions by:
CLAY STARK LOGAN
WALTER C. LOGAN II
ELMER KIDD
MARIAN KUNTZ
DICKSON TERRY

Photographs courtesy of:
THE STARK FAMILY
STARK BRO'S NURSERIES & ORCHARDS CO.
THE MISSOURI HISTORICAL SOCIETY
JIM LAW
JOHN STOECKLEY

Hardcover ISBN: 978-0-692-31291-9

This book is available in an e-book edition.
E-book ISBN: 978-0-692-31330-5

starkbros.com

Dedicated to Kyle David Brown, whose vision, imagination and embrace of technology set Stark Bro's Nurseries & Orchards Co. on a steady course for the next 200 years

Table of Contents

Foreword

In the fall of 1816, Abraham Lincoln was seven years old. Five years later, Missouri officially became a state. President John Adams would be alive for another decade when James Hart Stark, a stalwart Kentuckian and intrepid orchardist, brought his family and a saddlebag full of apple scions to set up camp near the mighty Mississippi.

Two hundred years later, Stark Bro's Nurseries & Orchards Co. still stands. This company — through a series of seemingly insurmountable travails — has become the world's oldest nursery, as well as world's largest online direct-to-consumer seller of fruit and nut trees, and berry plants.

There should be no mistake that the success of this company is due in part to good business sense, and in part to luck; but in the most part, because of the employees, the *people*, who have worked the fields, taken care of the customers and managed the business over these two centuries. We have a saying here: "Once you get moss down your shirt, you're always going to be at Stark." It's true; it gets in your blood and stays there. The work becomes part of you.

Sybil and Cameron Brown (center back), and their family

Many employees like to refer to themselves as the "other" Stark family, even though there are technically no longer any Stark family members employed here. They share strong bonds, forged through years of natural disasters, down financial times, births and deaths, and a common love of working the land. For this kind of dedication, there is no adequate adjective to express my appreciation. I hope the existence of this book will serve to do that.

On a personal note, I would like to add that my business partner, Tim Abair, and I share this caretaker mantle with our employees, as did my son, Kyle Brown. He was tragically killed in a car accident on December 14, 2011, but while he was here, provided the technological impetus for us to move into the next two hundred years. This book is dedicated to him by the author, a gesture for which I am grateful.

Cameron Brown

Cameron Brown
President and CEO
Stark Bro's Nurseries & Orchards Co.

1

Introduction

A Glossary of Terms

The Stark Bro's Nursery Operation

A Short History of Fruit-Growing in America

1 | INTRODUCTION

It survived the Civil War. It went through four financial panics, two World Wars and emerged from the Great Depression only to endure a 500-year flood, a fire and a couple of tornadoes. Not even the bankruptcy of its parent company could kill the 200-year-old Stark Bro's nursery, nor its spirit. The story of this company is one of resilience, pride, plenty of Midwestern pluck — and some plain old luck.

The first Stark Bro's office building, circa 1886

While many gardeners and commercial orchardists are familiar with the Stark name, the general public is not. What the public *is* familiar with are Stark's products. If you have Red Delicious, Golden Delicious or Gala apples in your refrigerator right now, they are there because of Stark Bro's Nurseries & Orchards Co. Surprised? So are most people.

The world's oldest continuously operating nursery is located in Louisiana, Missouri, situated on the Mississippi River just across the Illinois border. Missouri generally conjures up images of the grand St. Louis Arch, or the Vegas-style country music shows in Branson, but the landmark with the deepest roots — both figuratively and literally — is Stark Bro's (their chosen spelling). It is one of the country's oldest family businesses, having been started in 1816.

The men behind this remarkable enterprise were a formidable group of strong-willed personalities. The family business often crossed lines with the nursery business, and egos, money squabbles and bragging rights were all part of the equation. There were also moments of genius and displays of wisdom, leadership and kindness not often seen in the corporate world. With all of these attendant variables in play, it is a true wonder the company not only still stands, but thrives.

To think that there were no apples in America when the first colonists landed! Today, anything that is singularly typical of this country is said to be "as American as apple pie." The credit for that belongs largely to Stark Bro's.

To provide a sense of just how old this company really is, these are a few of the events that were transpiring in 1816:

- Napoleon had just been defeated at the Battle of Waterloo.
- James Monroe was elected president of the United States.
- Missouri would not be admitted to the Union as a state until five years later.
- The first savings bank in the U. S. opened.
- There was no running water, indoor plumbing, lights or heat.
- Thomas Jefferson was enjoying his retirement.

That's how long Stark Bro's Nurseries & Orchards Co. has been in business.

A Glossary of Terms

This is by no means a technical book, but there are several terms used throughout that are not usually found in the daily lexicon. These terms are explained more fully in later chapters.

Scion. A cutting or twig with several buds, or a single bud that is grafted onto a compatible rootstock. The scion is the above-ground or vegetative portion of the tree that becomes the trunk, branches, leaves and fruit. It is the part of the tree that has name recognition, such as Gala® or Honeycrisp™.

Rootstock. The underground part of the tree onto which the scion is grafted; this part of the tree provides the root/growth system for the tree.

Grafting. A skilled process which requires making slanted cuts on both the scion and the rootstock with a razor-sharp knife, then cutting a notch into the face of each sloping cut so that the two pieces fit. The pieces are taped together, eventually fuse and grow into a copy of the scion plant.

Budding. Much simpler than grafting, as it only requires one bud and its cambium layer to graft onto the rootstock.

Cultivar. A plant or tree that is cultivated by humans versus a variety that occurs in nature. Some cultivars originate as bud-sports (see below); other cultivars are a hybrid of two plants. To get a true clone of a cultivar, one must propagate asexually by grafting or budding. Trying to grow cultivars from seed almost alway produces something other than the true parent.

Bud-sport. A naturally-occurring branch or bud mutation that can be propagated and cloned to produce another whole specimen that is identical to its parent.

The Stark Bro's Nursery Operation

Stark Bro's has three main business areas: e-Commerce/Mail-Order, Commercial and the Garden Center. The operations include:

- 250 growing acres in Louisiana
- 427 growing acres in Atlas, IL
- 1 million cubic feet of climate-controlled warehouse space
- 1 million cubic feet of non-refrigerated work/storage area
- 1.7 million cubic feet of cold storage
- 5 acres of greenhouses
- 30,000 square feet of office space
- 1 million trees per year are shipped from Stark Bro's facilities

Aerial view of the Stark Bro's complex

What Stark Bro's Does

In addition to fruit trees — notably apple — Stark Bro's grows and sells nut and shade trees, berries, ornamentals and many other kinds of plants. Stark Bro's has created or discovered more than 300 superior varieties or strains of fruit. Depending upon the climate and soil conditions in their region, customers can grow and harvest anything from delicate apricots to exotic pawpaws, right in their own backyard.

The current Stark Bro's catalog offers more than 700 fruit, nut and landscape trees, and more than 250 berry and other garden plants, plus tools, supplies, specialty fertilizers, sprays and books. The Shipping Department sends out about 1 million trees per year, while the retail Garden Center sells many of the catalog items to walk-in retail customers.

The truth is, Stark doesn't grow trees "from scratch" any more. Since the early 1990s, they "build" trees, by using budding and grafting techniques vs. growing from seed. Visitors are sometimes disappointed because they come to Stark Bro's expecting to see rolling acres of apple trees in blossom, not realizing that the wonders Stark Bro's performs are with trees, not fruit. The fact is, the fruit is just a happy reward for their budding and grafting miracles.

DIY Landscaping

Stark Bro's was one of the first nurseries to introduce "do-it-yourself" landscaping. Long ago, the company recognized that hardly any of the home gardeners who bought Stark nursery stock really knew what to do with it once it arrived. The Starks brought in a cadre of young landscape architects — graduates of schools such as Cornell and Michigan State University — to set up a Landscaping Department and develop instructional booklets for their customers. These days, there is detailed information available online for every plant. Very knowledgeable customer service representatives also handle hundreds of customer questions about landscaping and horticulture every day.

The "old offices"

The Company Buildings, Old and New

The previous offices were across the way from the Garden Center, housed in a gorgeous Victorian mansion with a porte-cochère shading a sprawling lawn. It was built around 1900 as a home in case the nursery failed, according to John Logan, the company president from 1964 to 1984. Over the years, there were numerous additions, some of differing architectural styles. One of the additions, a turret, was used as executive offices for the head Starks.

The rambling, old painted-brick house section served as headquarters for an employee force of about 100 people to manage the mail orders. When the new office building was finished in 1979 (below), the old office was vacated.

By the time most of the quirky Victorian structure was razed (around 1982), the original core building's HVAC and electrical systems were out of code; the

second floor sagged a foot or more from one side to the other. Clay Stark Logan (John's son, and the last Stark descendant to be president) remembers a "rather large young lady who worked up there, and when she walked across the floor, it noticeably shook and seemed to bounce an inch or two."

Across the railroad tracks from the Garden Center is a five-acre tract of long packing houses, which are used in the fall, winter and early spring for protective storage and shipment. During these months, these sheds are full of trees, kept in large ricks (stacks) 12 to 25 feet high. The wonderful scent of a moist packing material called "shingletoe," which is made of cedar shavings, permeates the air.

Five acres' worth of well-kept greenhouses are filled with every kind of tree and plant imaginable, all meticulously tended by people who understand how to care for them. "Nursery" is a very apt term; millions of "babies" have been born there, where they're given everything they need until they can manage on their own. Then they begin their journey to the yard or orchard of a customer:

Order Processing

An order can be submitted to Stark Bro's via website, telephone, mail, fax or e-mail. When orders are placed, customers receive the best possible estimate of availability and ship dates for their location, based on the next ideal planting season (usually spring or fall). Once the customer has placed the order, they receive an e-mail confirming the details of the order.

Preparation

Order preparation consists of addressing any order adjustments, readying the item for shipment (pre-pruning, watering, bagging or boxing) and putting the product in the correct location for order fillers to easily find it.

Routing and Picking

Orders are released in the system one or two days before shipping. Pack-slips and batch sheets are printed and manually sorted each morning by an Operation Support Specialist based on order priority, and the product is "picked" in the warehouse.

The pack-slips and batch sheets are then taken to the Shipping Manager, who is responsible for making sure that each picking location supervisor distributes the information to the order fillers. A filler is responsible for pulling at least 10 orders at a time. Once the filler has loaded the cart, it is then spot-checked by an experienced checker to make sure all items have been correctly picked.

After the checker has cleared the cart to go, it is routed to the Packing Department, where it is boxed according to size, then stapled shut and manifested before it is placed on the truck. Trees arrive at the customer's door as fresh as when they left the nursery.

In prior years, the nursery used 1 million linear feet of lumber per year to make boxes; with today's lighter, stronger packaging materials, far less lumber is consumed. Solar panels and recycling efforts also conserve resources.

Shipment

When packing trees for shipment, the roots are carefully protected so that trees and plants arrive in "fresh-dug" condition.

To ensure reliable delivery of an order, Stark Bro's uses national delivery companies with solid reputations. Customers who have provided Stark Bro's with an e-mail address will receive an electronic shipping confirmation.

Stark Bro's trees and plants are fully guaranteed. If for any reason a customer is not completely satisfied, a replacement or refund will be issued within a year of purchase.

The Test Orchards

Since 1816, research has been an important part of the Stark operation. Today, the research work is considerably more complex, but it is directed toward the same end: the development of new and better fruits and the constant improvement of present strains.

Former VP of Production Jim Law, checking a record-bearing strain in the late 1960s

Among the more important work of the test orchards has been the development of a virus-free fruit-tree rootstock, production of disease-free ornamental trees and plants, methods of growing dwarf trees that adapt to a wider range of soil and climatic conditions, and disease-control methods in commercial orchards. Much of this work is carried on in cooperation with state, federal and international researchers. One day, the next "Delicious" may be discovered here!

There are now hundred of experiments taking place in the test orchards and experimental grounds, which are located in Louisiana, Missouri, and across the Mississippi River, in Atlas, Illinois.

A Short History of Fruit-Growing In America

It took a while for the lush fruits of the Mediterranean and Middle East to end up on our grocery-store shelves. Apples, pomegranates and figs are traceable back as far as 8,000 B.C., but didn't make it to these shores until the new American Colonies were founded in the early 17th century.

Down through the centuries more than 10,000 kinds of apple varieties have been developed. It was in England that selective breeding of apples first became a science, and the Jeniton apple was a prized discovery. Jeniton (Rall's Janet, also known as Ralls Genet) became a grower's favorite because of its flavor, texture and durability.

Although Americans are inclined to think of the apple as a typically American product, the first apples came from England with the first white settlers. They brought the famous Jeniton apples (and a great number of other fruits) across the Atlantic, and the seeds were literally sown for an American fruit-growing industry.

According to records from the Massachusetts Bay Company, apples were grown in New England as early as 1630. Nurseries and wholesale production primarily began in New York, Massachusetts and other east-coast states, starting with nutritious fruit trees and vines — the ornamental plants would have to wait, as vital food crops came first.

In 1737, the Prince Nursery, located in New York, became the first commercial nursery in the U. S. They introduced and developed many new fruit and rose varieties that are now household names in America.

As America was settled, traders, missionaries and Native Americans transported apple seeds westward. John Chapman (*aka* Johnny Appleseed) was responsible for extensive apple tree plantings in the Midwest in the very early 1800s. This benevolent apple-proselytizer wandered about the Ohio Valley, planting trees and freely distributing seeds to settlers heading westward in their Conestoga wagons.

Meanwhile, James Hart Stark and his family had staked out a plot of land in

Missouri, where he planted carefully-grafted apple trees (using Jeniton scions) that would become a regional sensation. That was the beginning of Stark Bro's Nurseries & Orchards Co., in 1816. As of this writing, Stark Bro's is the world's oldest continuously operating nursery.

Expansion of the nursery trade began booming after the construction of the Erie Canal was finished in 1825. In 1860, the advent of railroads significantly grew the nursery industry by reducing its sole reliance on waterway shipping. The railroads also opened up new markets by enabling salesmen and their nursery stock to travel quickly to customers.

In 1912, the U. S. Plant Quarantine Act and U. S. Parcel Post System powerfully impacted the fruit industry. The Plant Quarantine Act eliminated importing stock from European nurseries, and the Parcel Post System allowed nurseries to ship plants. This opened the door to national business for mail-order nurseries — including Stark Bro's, which eventually became the world's largest online retailer of direct-to-consumer nursery stock.

When the automobile came into widespread use in the 1920s, the industry would pivot again. People could now travel long distances in private vehicles, and so garden centers and mom-and-pop nurseries began to sprout up close to home.

What drives the fruit-growing business?

Housing sales and commercial development. When builders are building, trees are selling.

Popular culture. Designing with plants has been trending since the early 2000s; bold colors, textures, container gardening and urban street plantings (even urban farming) are very popular. Amid the debate on climate change, GMOs and chemical food additives, there is a movement to "grow your own" as the public becomes more interested in where their food comes from. As of 2014, The National Gardening Association says that 35 percent of all households in America are growing food at home or in a community garden, up 17 percent in five years.

An urban farm in downtown Chicago

Weather. Drought, heat, wind, flooding and unpredictable frosts all affect sales.

The economy and fuel prices. The more the economy is up, and the more gas prices are down, the better tree sales are. Disposable income in consumers' pockets is the key to the kingdom for nursery companies.

Nursery operations tend to fall into one or more of the following categories:

A field-production operation. These producers grow trees and shrubs in the ground that are then harvested for resale. The merchandise may be ball-and-burlap, pot-in-pot or bare-root, terms which refer to the way the item is harvested, prepared and shipped. A field-production operation usually includes a propagation nursery, like Stark Bro's has. Propagators (plant reproductionists) tend to be dedicated professionals and many are schooled horticulturists. A thorough grasp of the seasonal demand for the plants, as well as vigilant oversight of regional disease problems, are critical to success.

Pot-in-pot nurseries employ in-ground production with the built-in flexibility of container production. While the initial cost of pots and irrigation systems is high, over-wintering costs are not.

A pot-in-pot nursery

Container-plant producers grow plants in pots above-ground. When the plants reach a specified size, they are ready to sell. Stark Bro's uses a variation of this method, its proprietary 4-inch x 10-inch bottomless pot, a game-changing innovation and multi-award winner.

Specialty nurseries cater to growers with more exotic tastes, and to collectors. Most specialty operations sell via garden centers and websites.

Other nursery sub-types include:

Organic. To the USDA, certified organic means "agricultural products have been grown and processed according to specific standards of various state and private certification organizations." USDA-accredited agents review farm applications and qualified inspectors conduct annual on-site inspections. Farm records must track all operations, practices and materials used in their production, and a certified operation must have a written Organic Farm Plan available to the public. If all of these conditions are met, the USDA Organic seal may appear on organic apples and other 100-percent organic products, signs or advertising.

Pick-your-own. These farms allow customers to harvest fruit themselves, lessening the labor load for the farm: grading, storage and packing become non-existent. Such an operation requires "retail hours" and most of the business is concentrated on the weekends.

Online retailers. More than 86 percent of the U. S. population now has Internet access, and online reviews quickly help distinguish good companies from bad. In 1990, it would have been unthinkable to buy a fruit tree with your computer; now that sales channel is a measurable chunk of the $327 billion estimated to be spent online in 2016, according to the U. S. Online Retail Forecast.

The Nursery-Industry Outlook

Economies and taste-trends evolve, sometimes full-circle; ornamentals and specialty crops now reign over fruit. Big-box garden centers and independent landscape contractors have largely crowded out traditional family-run operations (with the exception of a few, like Stark Bro's).

According to *American Nurseryman Magazine*, since 1978 the low sales points are 1981-82, again in 1991, and then 2008-11. That pattern is not coincidental, if you ask Clay Stark Logan, the last member of the Stark family to run the company. His father, John Logan (president from 1964-1984), had graphed company sales for years, taping on a new piece of paper each year until there was a long, snake-like document that stretched across his desk.

"My dad's chart tracked Stark's sales and profits back to the early 1900s, and you can see a very clear up-and-down cycle about every 10 years," said Clay.

The ability to computerize operations and paperwork has drastically increased efficiency, and research has given growers better ways to irrigate, propagate and control weeds and pests. New cultivars are largely responsible for industry growth these days, which is why growers are so persistent about making and getting new discoveries through the research pipeline and into the marketplace.

The National Gardening Association says consumers spent $3.5 billion on food gardening in 2013 (vs. $2.5 billion in 2008), a 40-percent increase over five years — a good portent for the fruit-growing industry. The figures are a little deceiving because people are actually spending less per person, but there are more people spending, particularly those over 55 and under 34.

Where did apples come from?

The origin of the humble apple has been traced back as far as 8,000 B.C., as nomads from the Tien Shan mountains in China brought the fruit to the "civilized" Egyptian river valleys, where it began mutating into different varieties. Remains of apples dated to 6,500 B.C. were found among excavations at Jericho in the Jordan Valley.

Some historians believe the apple originated in the Dzungarian Alps, a mountain range separating Kazakhstan, Kyrgyzstan and China, where uncultivated trees still produce tiny, sour apples that barely resemble the Delicious and Gala varieties so popular today. Still others say the apple came from the Caucasus Mountains, between the Black and Caspian Seas. Whichever is true, it is clear that early nomads took these apples with them as they traveled to Asia, the Mediterranean and the Middle East.

In 323 B.C., the Greek botanist, Theophrastos, touted six different apple varieties, and was arguably the first to describe the art of budding and grafting techniques.

As Caesar's armies conquered continental Europe and the British Isles, they carried apple seeds and planted orchards to supplement native crabapples. By the first century A.D., orchard-grown apples were established as far north as the Rhine Valley of Germany. Later, apples were grown in monastery and convent orchards throughout Great Britain and Europe.

Dr. John Caius, 16th-century physician to three British monarchs, advised his patients to "smele to an old swete apple to recover strengthe." Perhaps this notion inspired this bit in the February 1866 edition of *Notes and Queries* magazine:

"Eat an apple on going to bed, and you'll keep the doctor from earning his bread." That little rhyme eventually morphed into, "An apple a day keeps the doctor away."

2

About Louisiana, Missouri

The James Hart Stark Cabin

The Stark Family Homes

The Stark Cemetery

2 | THE STARK FAMILY

About Louisiana, Missouri

Louisiana, Missouri is situated on the Mississippi River, about 90 miles north of St. Louis and 30 miles south of Hannibal. It is a city full of beautiful early-1800s homes, historic downtown buildings and a long stretch of majestic, old riverfront.

There is some dispute about the founding of Louisiana. The city's historical marker states that Louisiana's first settler was John Bryson, who arrived from South Carolina in 1816 to claim a quarter of land there. In 1817, he built the town's first residence on what is now Ninth Street, between Georgia Street and Tennessee Street. In 1818, Bryson sold some of the riverfront land to Joel Shaw and Samuel Caldwell of Kentucky, and this area became the original plat of Louisiana. *The History of Pike County* states that "the town was named for the state of Louisiana, and not after the beautiful face of some mythical Louise, as many have imagined."

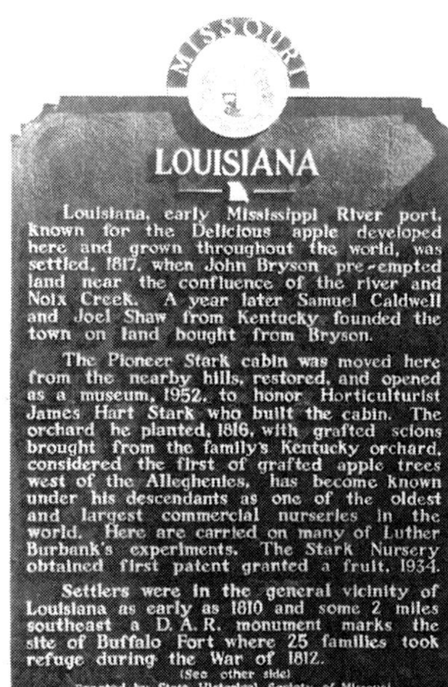

Others say the town was founded in 1817 by John Walter Basye (who also founded Bowling Green, Missouri). The story goes that it was named after his daughter, Louisiana, who was born in 1804, several months after the American flag replaced the territorial French flag, post-Louisiana Purchase.

Still others contend that the name was chosen to commemorate the Louisiana Purchase itself, and yet a fourth contingency insist that James Hart Stark and his band of settlers were there first, though he did not technically draw the plat or own the town center land, he had bragging rights for spotting the parcel and establishing a homestead there.

In any event, by 1835, Louisiana began to look like a real town, with leveled land, main streets and a levee to hold back the river. By 1837, the town had become an important trading post, with nearly 7,000 people who called it home.

Louisiana, Missouri is different now. In the 1980s, there were less than 4,500 residents, and as of 2013, just 3,355. Once the area's primary produce market, it was profoundly affected by the railroad expansion which made it easy to travel to the bigger, better markets of St. Louis and Chicago. Before, during and after the railroad's introduction, Louisiana remained dependent on the river. The long shipping barges still whistle a throwback to the days when Samuel Clemens (*aka* Mark Twain) piloted steamboats along the Mississippi, past Pike County.

At the same time that Bryson, Shaw and Caldwell were developing the town of Louisiana near the river, James Hart Stark was starting a nursery at the western end of town. From this humble beginning grew Stark Bro's Nurseries & Orchards Co., still a flourishing business, and still located in Louisiana. It is now the largest online direct-to-consumer nursery in the world.

The James Hart Stark Cabin

While Louisiana, Missouri is a small rural town absent the glitter of more cosmopolitan locales, the "Stark mansions" that dot the landscape are plenty grand enough. The family's beginnings, however, were definitely humble.

The Stark Cabin: Hwy NN at Business US 54/Georgia Street

James Stark built the original Stark Cabin upon his arrival in Buffalo Township in 1816. His young family and grandson Edgar were all raised up there.

The cabin was moved to its present site and restored as a museum in 1951. As it was dismantled in preparation for the move, each piece was numbered and then reassembled, just as it stood in Buffalo Township — down to the square handmade nails in the floor. The only change was to put cement between the logs instead of the old chink-and-daub plaster. It was then furnished with family artifacts and early- to mid-1800s antiques, and periodically opened to the public.

When the cabin was open for tours, one could see James Hart Stark's old Kentucky long rifle perched over the entrance door. The furniture was primarily from his son William's family. The Blue Willow ware in the hutch was from Belgium. Around the fireplace, a treasure trove: a warming oven that reflected heat into the room; a broadaxe blade; a cast-iron beetle boot-jack, an old froe (wood-splitter), sadirons for pressing clothes, a melting pot for making bullets and a fire-lighter pot, all essential items in pioneer times.

The pie safe in the opposite corner displayed a number of the hand-whittled utensils used by the Stark family. On the mantle shelf sat a foot-warmer, a little jug-like affair made for holding hot water to put in the bed or in the sleigh. Apple-butter stirring paddles, a fluting iron and a Civil War-era oil lamp are a few of the other items that portrayed life at that time.

William Stark's rope-slat bed was here, replete with a trundle for the young children. Rare and beautiful antiques like cobalt glass and a Flow Blue pitcher-and-bowl set seemed almost out of place in this humble cabin.

The child's chair was of special interest; that belonged to Edgar, one of William's sons, the third of six generations to run Stark Bro's. He grew up in the cabin and later lived at 1905 Georgia Street, closer to the town.

The tours stopped after an arsonist scorched the inside of the building in the late 1960s and destroyed many of the irreplaceable antiques. The last cabin tour was given in 1991 at the 175th anniversary celebration. In September of 2011, the cabin got a new hand-hewn, cedar-shake roof and is still standing strong. James Hart Stark's first Missouri home serves as a daily reminder for every employee who drives past it why they have a job.

The Stark Family Homes

They say "money doesn't grow on trees," but it can assuredly grow from selling trees. The Starks built luxurious residences all around the Louisiana area, queen among them the Gov. Lloyd Crow Stark home, a two-story Eastlake mansion with an adjoining carriage house (1401 Georgia Street). Built in 1891, the buildings remain substantially unchanged from their days as the Stark family residence between 1915 and 1936. Governor Stark also lived at 1210 Georgia Street and the Aberdeen Farm a few miles north of Eolia, Missouri.

Clay and Walter Logan, the last two Starks to run the company, grew up in the house at 201 Noyes Street. Walter lived down the way at 210 Noyes Street as a young adult, and moved to 1905 Georgia Street in 1987.

The home of Governor Lloyd Crow Stark

Other notable Stark homes and residents, past and present:

Paul C. Stark Sr.	320 N. Main Street
Paul C. Stark Jr.	819 Kentucky Street
	1111 Georgia Street
	901 N. Third Street
Paul C. Stark III	1000 Georgia Street
Edgar W. Stark	1905 Georgia Street (built by Edgar)
	The Stark Cabin
Clay H. Stark	1905 Georgia Street
Lloyd "Stickney" Stark Sr.	3210 Georgia Street
Lloyd "Stickney" Stark Jr.	321 N. Fifth Street
Edwin J. Stark	417 N. Main Street
Lawrence E. Stark	220 N. Third Street
John Stark Logan	201 Noyes Street
Walter C. Logan, II	210 Noyes Street
	1905 Georgia Street
Clay Stark Logan	417 N. Main Street
Thomas "Tom" W. Stark	1401 Georgia Street
Thornton Grimsley Stark	1114 Georgia Street
Clarence M. Stark	601 N. Third Street (built by Clarence, *aka* the Goodman-Stark House)
Willela Stark Unsell	601 N. Third Street (Clarence's daughter and son-in-law)
James Ovid Stark	14944 Hwy D ("The Stark Mansion")
Eugene W. Stark	14944 Hwy D

The Stark Cemetery

The final home, as it were, of many of the Starks is about nine miles away from company headquarters, in the unassuming Stark Cemetery. The family resting place is situated in the pastoral rolling hills of northwest Calumet Township on Highway D, about one mile north of Calumet. There is neither a physical map nor a record of the plots, but many of the Stark headstones were transcribed by Homer and Sally Sterne, J. D. Burns and Rosemary Brown in 1984. Several of the headstones have fallen or broken, which accounts for some of the missing dates below.

Grave marker of Judge James Hart Stark in the Stark Cemetery

Stark, John:	2/27/1794 - 9/13/1867
Stark, James Hart:	7/30/1792 - 5/27/1873
Stark, Henry:	3/19/1816 - 6/9/1840
Stark, Thomas:	4/17/1817 - 11/1/1849
Stark, John W.:	7/28/1818 - 7/24/1872
Stark, John B.:	9/4/1851 - 8/1875
Stark, James E.:	11/10/1871 - 9/15/1875
Stark, Thornton G:	1830 - 1882
Stark, Jane Watts:	8/22/1797 - 12/29/1882
Stark, Elizabeth:	6/2/1822 - 12/31/1899
Jump, Susanna Stark:	8/14/1823 - 2/28/1907
Stark, E. Homer:	1858 - 1915
Stark, Lucinda E:	1829 - 1919
Stark, James Ovid:	1855 - 1927

There are also a number of Starks buried in Riverview Cemetery in Louisiana, Missouri, including Clay Hamilton Stark, Clarence McDowell Stark and Paul C. Stark III.

Grave marker of company president Clay Hamilton Stark in Riverview Cemetery

3

The Muirhead/Starks of Scotland

The Starks Come To America

From Virginia to Kentucky

The Stark Family Tree

3 | THE ORIGIN OF THE LINEAGE

The Muirhead/Starks of Scotland

The very first Stark was actually born John Muirhead, who grew up to be the bodyguard of King James IV of Scotland.

While hunting one day in 1642, he saw that an angry bull was about to gore the King, who was standing under a copse of oak trees. With no care for his own safety, young John lept forward and grabbed the bull's horns, tearing the beast away and saving the monarch.

The Stark Family crest

In gratitude, King James decreed that John Muirhead and his descendants should henceforth bear the name of Stark, meaning "strength." The family crest was ordered to bear a bull's head and three acorns, symbolizing the oak trees under which the King had been standing. The motto would be this: *Fortiorum Fortia Facta* ("Brave Deeds of Brave Men").

The newly-renamed Stark clan, whose home was in Dunbarton, was also granted land in the Cumberland area of Scotland, near Loch Lomond. For almost two centuries they lived and flourished at Killermont estate.

Killermont Estate, circa 1628, near Glasgow, Scotland. The Starks sold Killermont in 1685. Since 1904, it has been a golf country club.

The Starks Come To America

The first John Stark to come "across the pond" emigrated to New Hampshire in 1710, and was later joined by his four children. As a protection from the native Indians, he built a stockade that became known as Stark's Fort. The eldest of the sons, in a break from family tradition, had been named Archibald (instead of the usual John), for his mother's family. Archibald raised his children in America; one of his sons was to become the famous General John Stark, hero of the Battle of Bennington during the Revolutionary War. He is credited with coining the New Hampshire state motto: "Live Free Or Die."

As other migrants came to Stark's Fort, the settlement became known as Starkstown, and then Dunbarton, for the old Stark hometown in Scotland.

James Stark, second oldest of John Stark's children, was 20 years old in 1716 when he married Elizabeth Thornton, of Welsh descent. (Her brother, Dr. Matthew Thornton, would some 60 years later be a signer of the Declaration of Independence.) James and Elizabeth had eight children, the eldest named John. In 1730, they moved to Stafford County, Virginia, where seven more children were born to them.

General John Stark

In 1746 James' eldest, John, married Hanson Porter, who was from an old Virginia family noted for its naval officers and ship's captains. After 30 years of marriage, John became a widower. His second wedding was to Hanna Eaves, his children's fetching Welsh nanny; they had seven children of their own. John was an attorney during the week, a preacher on Sundays and a courageous colonel in the French and Indian Wars.

This statue was given to the National Statuary Hall Collection by New Hampshire in 1894.

John's oldest son by his second marriage, James, was born in 1757. Like all of the Johns and Jameses before him, he was filled with indefatigable energy. During the Revolutionary War, when he was only 14, he saddled his horse, took his gun and rode off without a word to anyone. He sought out his father and his father's cousin, General John Stark, who were fighting with the Americans. He found them preparing for a battle.

Needless to say, the soldiers were alarmed at the prospect of having such a young boy in harm's presence, but they eventually realized that he had done the only thing a Stark could do: be brave. They gave him a place in the ranks. After the Boston Tea Party in 1773 he was made a captain, after which he admirably served with his famous kinsman General Stark at the Battle of Bennington.

Three years later, the Captain was married in Stafford County, Virginia, to Miss Jane Fristoe. She died in 1785, leaving him with two small daughters,

General John Stark

General Stark's great victory at Bennington not only assured his place in history, but his wife's as well. Every American is familiar with the famous admonition of war, delivered on the eve of the battle in 1877: "There are the red coats and they are ours, or this night Molly Stark sleeps a widow."

New York State historical marker, erected in 1929

During the years after the close of his military life, General Stark devoted himself successfully to business. Many — including George Washington, Thomas Jefferson and James Madison — expressed their appreciation of his services to his country in the years after the war.

another daughter having died in infancy. Later that year, Captain Stark was given a grant of some 2,000 acres in Bourbon County, Kentucky, a reward for his military service.

From Virginia to Kentucky

And so Captain James and his two daughters, along with brother Thomas Stark and his wife, set out for Kentucky in 1785 to homestead his land grant. They were guided by the famous frontiersman Daniel Boone. The route was the Wilderness Road, a treacherous 600-mile trail from the western border of Virginia, through Tennessee's Cumberland Gap, then west to Fayette County, Kentucky. The road was rife with danger from the notoriously savage Chickamauga tribe of Tennessee, as well as the Shawnees and Mingos from northern Ohio;. Even after settling in Kentucky, both James' and Thomas' families were frequently forced to seek protection at Bryant Station, a fortified settlement nearby.

A year after arriving in Kentucky, James Stark married Susanna Hart, whose family was from Kinlock, Scotland. They had eight children. Presumably Captain Stark led the life of a typical country squire in Kentucky, tending extensive orchards grown from scions taken from his father's Virginia fields.

The fourth child by his second marriage, born in 1792, was James Hart Stark, who was destined to one day cross the Mississippi River and begin a dynasty.

THE STARK

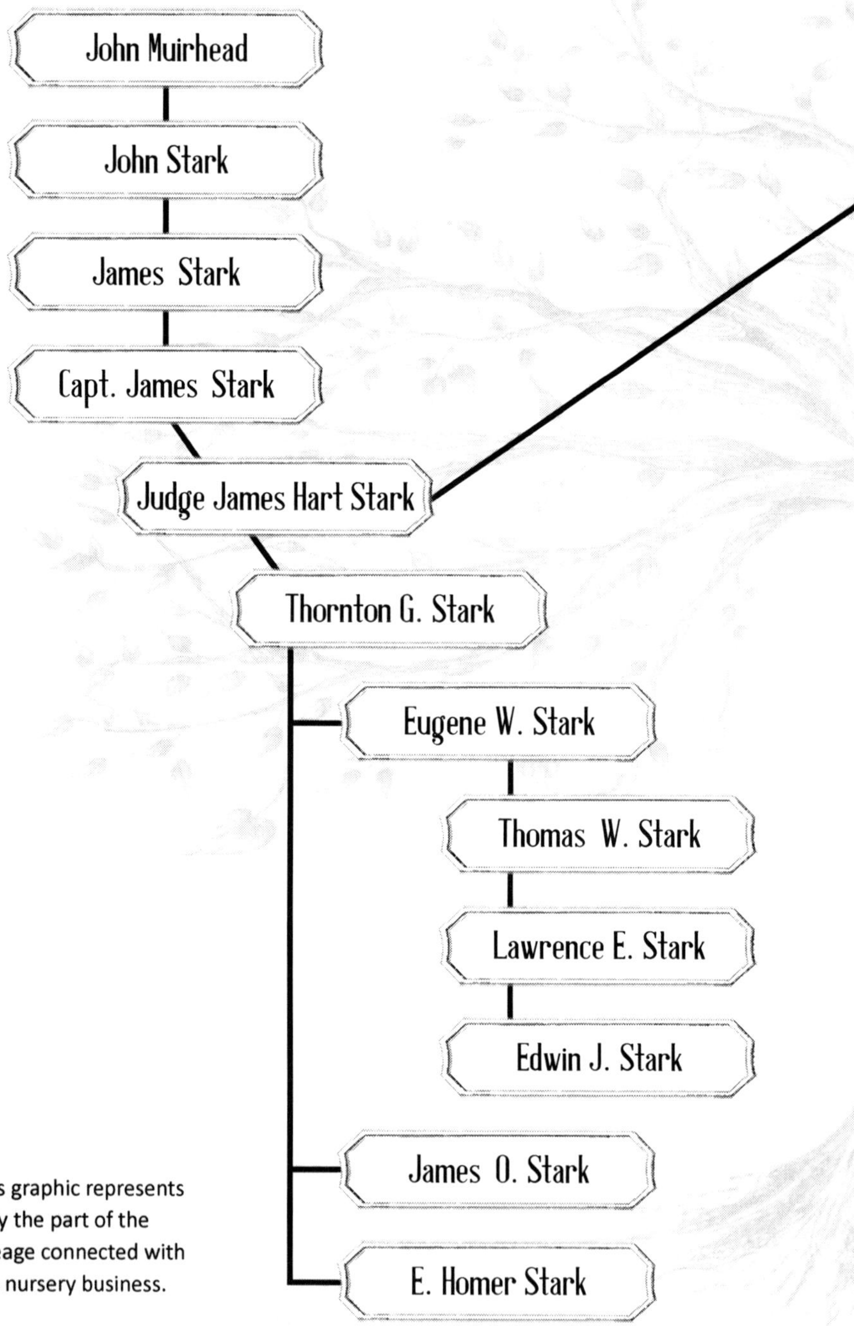

This graphic represents only the part of the lineage connected with the nursery business.

FAMILY TREE

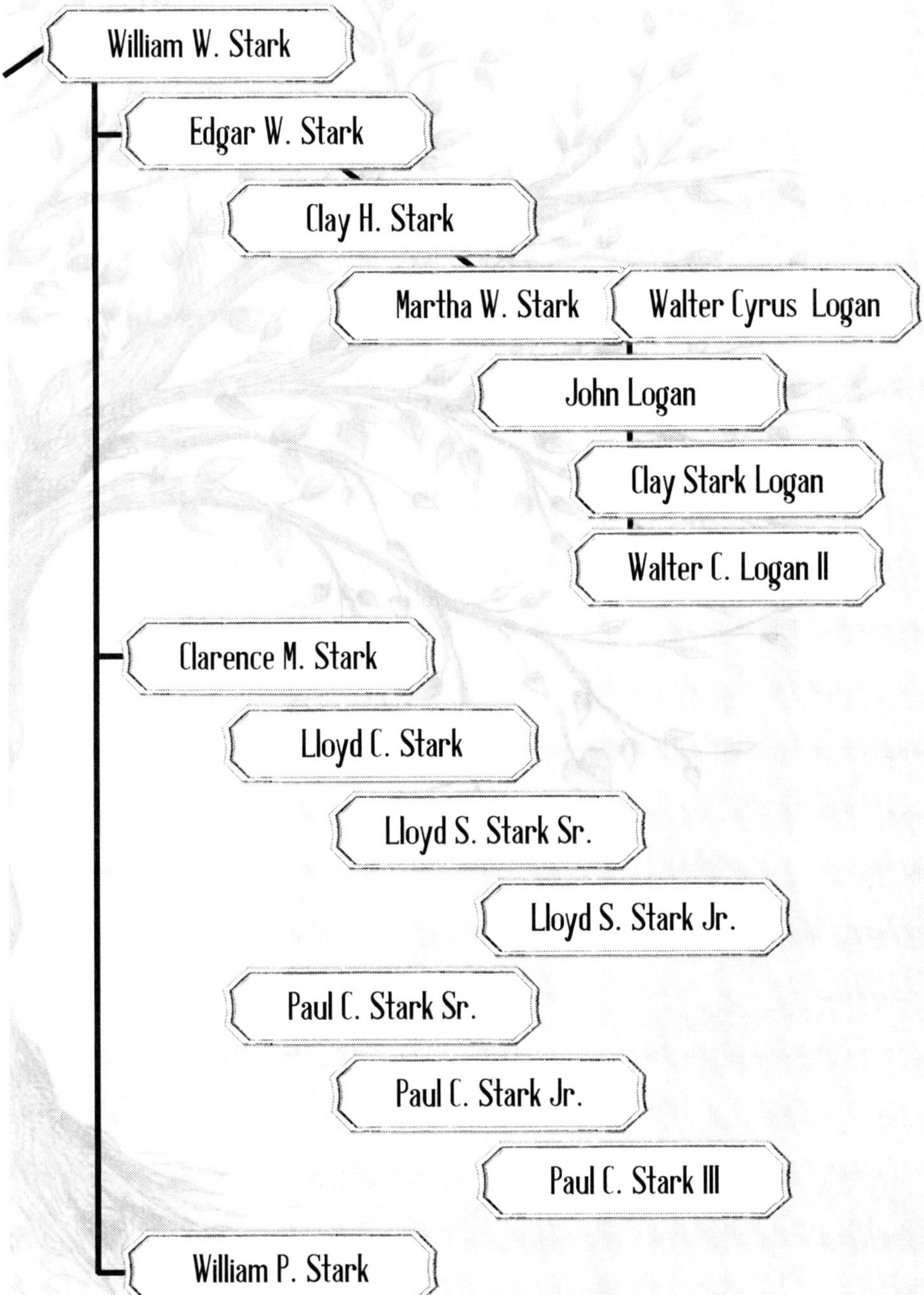
William W. Stark
Edgar W. Stark
Clay H. Stark
Martha W. Stark
Walter Cyrus Logan
John Logan
Clay Stark Logan
Walter C. Logan II
Clarence M. Stark
Lloyd C. Stark
Lloyd S. Stark Sr.
Lloyd S. Stark Jr.
Paul C. Stark Sr.
Paul C. Stark Jr.
Paul C. Stark III
William P. Stark

4

Biographies of the Stark Family Nurserymen:

WIlliam Watts Stark Branch

Thornton Grimsley Branch

Biographies of the The Current Owners

4 | THE STARK NURSERYMEN

Stark Family Members

(Titles and years of service refer to the last position held)

JAMES HART STARK (1792 — 1873) / First Generation

President, 1816 - 1855

JAMES HART STARK

The rather stern-faced founder of Stark Bro's Nurseries was born in 1792 in Bourbon County, Kentucky, to Captain James Stark and his second wife, Susanna Hart. From a capable orchardist at 14 to a brave infantryman at just 16, he fought in the Tippecanoe and Thames battles and came home a captain.

In 1815 he married Jane Watts with the intent of settling down, but had heard about the Great West and its promises of free, fertile land. A band of 20 men made an exploratory trip to Pike County, Missouri in 1815, and a year later James returned with his family and erected a small settlement on the banks of the Mississippi. He had brought apple scions from his father's orchard in Kentucky and began grafting them onto the native crabapple rootstock, unwittingly launching a horticultural empire.

James ultimately fathered 17 children, only one of whom would succeed him in the nursery business (William Watts Stark). From 1840 to 1844 James served as Pike County's tough-but-fair judge, an occupation befitting his stern and honest nature. James Hart Stark died of heart disease on May 27, 1873.

His work, and the work of those who came after him, brought worldwide notice to Missouri as well as to the field of horticulture. He was an early practitioner of scientific plant breeding, which produced superior strains of fruit. Without him, there would likely be no Red Delicious, Golden Delicious or Gala apples, as these were all discovered and cultivated by his descendants.

❖❖❖ THE WILLIAM WATTS STARK BRANCH ❖❖❖

WILLIAM WATTS STARK (1827 — 1880) / Second Generation

President, 1855 - 1880

WILLIAM WATTS STARK

Of Judge James Stark's 17 children, only one of his nine sons, William, chose to carry on the nursery business. The judge's youngest son, Washington, worked with his father until 1857, but then married and moved away to Pettis County, Missouri where he bought land parcels to start his own nursery venture. The enterprise lasted but a few short years — it was unceremoniously ended by the onset of the Civil War.

William was a natural horticulturist who learned budding, grafting and selection by working along with his father when still a young child. There were no schools of horticulture in those days, so William read everything he could find on the subject. He succeeded his father at the age of just 14.

He married Cynthia Eliza Pharr in 1853; she would be the mother of his five children. By 1855, William's son Thornton joined the operation.

William and a close friend organized the first Missouri State Horticultural Society meeting in 1859 and he served as vice-president for 12 years thereafter. His yearly reports were full of valuable wisdom that still holds up today; his wealth of knowledge was based on real-world experience and a treasure-trove of collected insights from the other nurserymen in his family. The advice he gave was a veritable how-to for orchardists everywhere; William was recognized as an authority and was considered a sort of horticultural rock star, if you will.

While so many of the Starks served in the military, William did not, due to a heart condition. Instead, he served in law enforcement as the deputy sheriff of Pike County, and later replaced the sheriff, who was killed by local fighters.

When the railroad was completed in 1870 through Noix Creek Valley, William moved the nursery from its original spot in Buffalo Township to its present location in Stark Park, directly across the way from the Garden Center and outbuildings, where he would work for another 10 years.

After the war, William bought his brother out, and by the early 1870s he was the beneficiary of an admirable enterprise, at least for a short while. The financial

Panic of 1873 and that year's harsh winter took a severe toll on the business, and William was never the same after that. It was at this time that William's three sons — Clarence McDowell Stark, Edgar Winfred Stark and William Pharr Stark — joined the business as naïve young men with little agricultural knowledge, but nonetheless assumed the operations of the business upon their father's death in 1880.

CLARENCE McDOWELL STARK (1855 — 1914) / Third Generation

President, 1880 - circa 1906, officially 1911

CLARENCE McDOWELL STARK

Clarence was the oldest of William W.'s sons. In high school, he was a student of Champ Clark, the famous House Speaker, who became a family friend. In 1885 Clarence married Lilly Crow, the daughter of former Pike County folks who had moved west to California. Their children were Lloyd Crow, Willela and Paul C. Stark Sr.

It was Clarence who proclaimed an 1893 apple sample from an Iowa farmer to be "delicious," thereby launching what became a household name in fruit.

Clarence was said to have tuberculosis (though it was actually likely Alzheimer's or dementia) in 1887, and temporarily moved to Colorado as a "cure." He recovered enough to incorporate the company as Stark Bro's Co. in 1889, and, after the financial Panic of 1893 rightly realized that advertising in down-times was key to keeping the business afloat. He instigated the company's flagship catalog; his marketing acumen was legendary.

His mental health continued to deteriorate, and by at least 1906 he was in a Texas sanitarium. In 1911, he was declared insane by a San Antonio probate court. After that, his wife Lilly moved him to Maplewood, another sanitarium in Jacksonville, Illinois, with the court's permission.

He died in 1914 as an involuntary patient in that sanitarium, before he could see the great success of the Golden Delicious apple fully realized. A coroner's inquest determined that he died of a self-administered dose of anxiety and sleep medication.

PAUL CLARENCE STARK SR. (1891 — 1974) / Fourth Generation

VP and General Manager, 1935 - 1974

PAUL CLARENCE STARK SR.

Paul, Clarence's youngest son, began his nursery career at the age of six. He grew up to attend Cornell University and obtained a degree in horticulture in 1912. He was responsible for discovering the hybrid apple known as Golden Delicious, which remains one of the most popular apples in the world. He is also credited with the creation of the Starking® Delicious apple, an earlier red apple which revolutionized the nursery industry.

Paul married Theodosia Barnett in 1916. He became company Treasurer in 1919 and was eventually made VP and General Manager in 1935 when his uncle Edgar passed away. He was awarded the prestigious Wilder Medal by the American Pomological Society in 1972 for his discoveries, and eventually became the society's president. He died in 1974 of a heart attack at a St. Louis hospital.

PAUL C. STARK JR. 1917 — 1989 / Fifth Generation

VP for Special Research, 1967 - 1982

PAUL CLARENCE STARK JR.

At the age of just two, Paul Jr.'s photo appeared on the cover of the 1919 Fruit Planning Guide, holding a Stark Delicious apple. In 1940, he graduated from Cornell University with a B. S. degree in pomology. He was an instructor in fruit science from 1940-1941 at the University of California-Davis, then came back to Missouri to work at the nursery, starting a 41-year career full of accomplishments that would change the way we grow apples. He was very much like his father: inquisitive, bright and forward-thinking, but he did have difficulties: Paul was married four times (to Janet, Wilma, Hadley and Lois) and was known to be overly fond of alcohol. Personal money management issues also plagued him.

His career was interrupted while he took a four-year tour of duty in World

War II as a U. S. Navy Senior Commissary Officer. Upon his return, he would begin traveling the world in search of new varieties. In 1967, he was promoted to VP for Special Research in the field of breeding, selection and propagation of deciduous fruits, principally apples. He was an important influence in the field, working closely with many other widely-known horticulturists. Paul Jr. was noted for his keen ability to spot new varieties and interpret their probable value to commercial growers.

Spur-type apple cultivars like Starkrimson® Red Delicious ('Bisbee') and Starkspur® Golden Delicious (Gilbert strain) were discovered by Paul Stark Jr., which introduced spur-type semi-dwarf trees to the industry, changing it forever.

He was president of the Missouri State Horticultural Society; a long-time member of the American Pomological Society; received the American Pomological Society Award in 1960 for his paper on "Fruit Growing in the South Pacific," the Home Orchard Society Award (1978) and the International Dwarf Fruit Tree Association Award (1975).

Paul Stark Jr. passed away at the age of 72 at Pike County Memorial Hospital, where he had been a patient for three weeks. His body was donated to the University of Missouri Medical School in Columbia.

PAUL C. STARK III ("PIII") 1949 — 2004 / Sixth Generation

Assistant Secretary, 1975 - 1985 (dates are approximate)

PAUL CLARENCE STARK III

"PIII," as he was known, was born in St. Louis in 1949 to Janet Merrill and Paul Stark Jr. He was raised in Louisiana, Missouri but later moved to California, where he earned a B. S. degree from the University of California at Davis.

Paul Stark III was, by many accounts, the last Stark with a research background in agriculture to work in the family business. In the mid-'70s, Paul came to work at Stark Bro's, but left about nine years later to go to California and "pursue other things." He served with the U. S. Naval Reserves from 1971-1977, was a member of St. John's Episcopal Church and sang in a barbershop quartet. Sadly, he died in Santa Cruz, California at 55 of a brain aneurysm. PIII's 27-year-old daughter, Jeanette, died several months later of unknown causes, only compounding the tragedy. Paul Stark III also left another daughter, Laura. He is buried in Riverview Cemetery.

LLOYD CROW STARK (1887 — 1972) / Fourth Generation

VP and General Manager, 1912 - 1935

Lloyd Crow Stark was born and raised in Louisiana, and was Paul Sr.'s older brother. After graduating from the United States Naval Academy in 1908, he married Margaret Stickney, with whom he had four children. Lloyd served as a Major in the United States Army from 1908 to 1912. Upon returning, he went to work with the family business as VP, but his career was interrupted by World War I when he volunteered for Army service again. After he returned the second time, Lloyd expanded the business and profits quadrupled. By the 1930s, more than 22,000 people worked for the nursery and Lloyd became one of the most influential businessmen in Missouri.

He had also developed an interest in politics, business and city planning, and first entered that arena when he was appointed chairman of the Citizens' Road Bond Committee in 1928 by then-Governor Baker. In that capacity, he worked for passage of a $75,000,000 bond issue that would ultimately establish the farm-to-market road system. Lloyd was instrumental in making the highway bridge over the Mississippi River at Louisiana a reality. His first wife, Margaret, died in 1930, and he remarried in 1931 to Katherine Perkins. A Democrat, Stark was elected Governor in 1936, and ran for the U. S. Senate in 1940, but lost in the primary to Harry Truman by only 8,000 votes. Stark was largely credited with the destruction of the powerful political machine of Tom Pendergast in Kansas City in the 1930s and 1940s. This defeat signaled the end of Stark's own political career, and he thereafter devoted himself to the management of Stark Bro's Nurseries. He died in 1972 of natural causes, but had suffered a stroke not long before he passed away.

LLOYD STICKNEY STARK SR. (1908 — 1946) / Fifth Generation

Assistant Secretary, 1935 - 1939

Lloyd Stickney Stark Jr., Lloyd Crow's oldest son, studied Landscape Gardening and Horticulture at

Cornell, and Business Administration at Washington University. He resigned from his position in the company in 1939 to accept a commission in the United States Navy. He eventually became a Lieutenant Commander, and was on duty at Pearl Harbor when the U. S. fleet was attacked. There he helped save many men in America's overturned battleships. After Pearl Harbor, he was ordered to Mare Island Navy Yard as a Security Officer. When the Navy yard at Hunter's Point, San Francisco, was completed, he became a Senior Security Officer. While there, he received a secret shipment (the first two atomic bombs) which he had loaded on a cruiser. The bombs were shipped to Tinian Island, where Governor Lloyd Stark's Annapolis roommate for four years, Admiral Purnell, assembled them for the bomber that dropped them and ended the second World War. Lieutenant Commander Stickney Stark was tragically killed in a C-27 transport crash on his way to a family reunion in Missouri. He was 38.

JOHN WINGATE STARK

JOHN WINGATE STARK
(1916 — 1987) / Fifth Generation

John was the second son of Lloyd C. Stark. In 1940 he enlisted and became a lieutenant in the Army Air Corps. He returned to the nursery for a time, working in the photographic and catalog department, but soon resigned and went back into the military. He served again as a Major in the Air Corps Reserve, and eventually received his Master's degree in Education from the University of Arizona.

He died in 1987 and is buried in Tucson, Arizona.

LLOYD STICKNEY STARK JR.

LLOYD STICKNEY STARK JR.
(1930 — 2015) / Sixth Generation

Vice President of Marketing, 1972 - 1974

Lloyd Stickney Stark Jr. served as Lieutenant, U. S. Army, in Germany. He was at one time vice-president of Stark Nurseries, but resigned in 1974.

He married Frances Fulda, a Mexican national, in 1953. Their children are James Lloyd (born 1954) and Frances Elizabeth (born 1957). His second marriage was to Susan Clayton Rendlen; they divorced after around 10 years at which time Lloyd left Stark Bro's and moved to Oregon, where he married Lamyra French. Lloyd Jr. passed away in Mexico in 2015.

EDGAR WINFRED STARK (1856 — 1935) / Third Generation

President, 1914-1935

Edgar was William W.'s middle son: stable, hard-working and forward-thinking. He married Martha "Mattie" Johnston in 1887 and they had two children, Martha and Clay.

Edgar wore many hats at the nursery: he was in charge of production, shipping, real estate and finance, and was known as the company's "efficiency expert."

In tandem with Clarence, Edgar smartly laid the foundation for expanding the business nationally and overseas; with the great Homer Reed, he developed the mechanical tree digger and brought many outdoor operations indoors, yet another Stark to revolutionize the industry.

He died of atherosclerosis in 1935 at the age of 78, after spending 60 years in the nursery business.

CLAY HAMILTON STARK (1894 — 1953) / Fourth Generation

President, 1935 - 1953 (died in office)

Clay Hamilton Stark was born in 1894 in Pike County. His parents were Martha "Mattie" Johnston and Edgar Winfred Stark. Clay's sister was Martha Winfred Stark, who married Walter Cyrus Logan, of Hannibal, Missouri. The Stark/Logans (namely their son John, and grandsons Clay and Walter II) would come to be the last Starks to be in charge of the nursery business.

Clay attended school in Louisiana and graduated from a military prep school in Tennessee. He attended Washington and Lee University in

Virginia for one year, then transferred to the University of Missouri at Columbia. During his senior year, he joined the French Ambulance Corps and did indeed drive an ambulance in France. When the United States entered World War I, he transferred to the American Expeditionary Forces and served in the tank corps of the United States Army.

In 1919, Clay joined Stark Bro's, the fourth generation of the family to do so, and took over as president when his father Edgar died in 1935. He married Elizabeth Harris, daughter of Missouri's Lieutenant Governor, in 1922.

Clay Hamilton Stark died in 1953 of lung cancer at the age of 58.

JOHN STARK LOGAN (1918 — 1995) / Fifth Generation

President, 1964 - 1984

The Logan name was introduced to the family through John Logan's mother, Martha Winfred Stark, who had married Walter Cyrus Logan. John, who was born in Hannibal, Missouri, was the grandson of Edgar Winfred Stark. He attended Summer Naval School at Culver Military Academy from 1932-1934, then the University of Missouri, receiving a degree in business and public administration. (Sam Walton, a classmate, later founded Walmart.) Beginning in 1935, he worked part-time for Stark Bro's while living with his grandparents in Louisiana.

While at college, John was the president of his fraternity and managed the football team in the 1939 Orange Bowl. He enlisted in the Navy in 1940, only to be discharged a month later due to color-blindness. In 1941 he was drafted into the Army, served in Ireland and was honorably discharged as a Captain in the Army Medical Corp. John married Mary Alice Woodward in 1947.

He started full-time employment at Stark Bro's in 1945, and became VP, Assistant Secretary and Assistant Treasurer in 1953, then president in 1964. John served as the VP of the American Association of Nurserymen while with the nursery, as well as the president of the National Mail Order Nurserymen's Association (now the Direct Gardening Association). Additionally, he served as president of the National Association of Plant Patent Owners from 1974-75.

John was very civic-minded, serving on the school board, in the Chamber of Commerce (local and state) and on the Board of Directors of the Bank of Louisiana and the local United Fund. He was active in many fraternal and service organizations, receiving the University of Missouri's Outstanding Alumnus Award in 1971 and the Louisiana, Missouri Citizen of the Year by the Chamber of Commerce in 1980. He was inducted into the Mail Order Association of Nurserymen Hall of Fame and appeared on the popular television show "To Tell The Truth" in 1966 as president of the "Largest Nursery in the World." He retired when his son Clay assumed the presidency in 1984, but remained a valuable unpaid consultant until his death in 1995.

CLAY STARK LOGAN (1948 —) / (Sixth Generation)

President, 1984 - 1999

Clay, John's older son, attended Westminster College in Fulton, Missouri, and joined the service-oriented Phi Gamma Delta fraternity.

He was also an Army ROTC cadet at Westminster and was commissioned as a second lieutenant upon graduating in 1970. He received a medical discharge from the Army prior to deployment to Vietnam.

Clay's nursery career began in 1972, as a trainee in Inventory. From there, he became the nursery stock inventory manager, and then transitioned to Assistant to the President (his father). In 1978, Clay married Sally, and they have a son, John.

In 1978-79 he oversaw the construction of the new corporate office building, and when the recession hit, he helped consolidate the branch locations and diversify the business. As his father neared retirement, Clay was virtually serving in the president's capacity, but was not officially given the title until 1984. He held that position until the business was sold in 1994, and like his brother Walter, remained for the next five years on contract with the new owners.

Clay's list of civic and business activities is extensive. Representative of the positions Clay has held include: Pike County Memorial Hospital (PCMH) Trustee, President and Treasurer; Bank of Louisiana Director; Rotary Club of Louisiana Member, Director and President; Rotary District Assistant Governor; Louisiana, Missouri City Councilperson and First Presbyterian Church Elder.

He has also served as Missouri Association of Nurserymen Board Member and President; Mail Order Association of Missouri Board Member and

President; National Association of Plant Patent Owners Director and President; University of Missouri Horticulture Farm Advisory Council Member; Missouri Chamber of Commerce, the Cub Scouts, local schools and too many others to list. In 2013, Missouri Community Betterment awarded Clay an Adult Leadership Award for his dedication to city and community matters.

WALTER CYRUS LOGAN II (1952 —) / (Sixth Generation)

VP and Sales Manager, 1983 - 1999

WALTER CYRUS LOGAN II

Walter is the younger son of John Logan. He attended Westminster College in Fulton, Missouri, where he was a member of Phi Gamma Delta social fraternity, serving as treasurer for two years. He was an ROTC cadet who was honorably discharged due to allergies and asthma. He graduated with a degree in economics and business administration in 1974.

Walter moved to Fort Worth, Texas to be with his fiancée, Cindy, whom he married in 1975. Cindy and Walter have three children: Benjamin John, Bethany Grace and Michael "Blake."

After a two-year stint with Miller Brewing Company's accounting department, he and Cindy returned to Louisiana, Missouri and Walter joined the family business. After several promotions, Walter became VP/Mail Order Sales Manager and held that position from 1983 until the sale of the company in 1994. He stayed on for five additional years as VP of Marketing, then moved to Indiana as Director of Marketing for Gardens Alive! for one year. In 2001, Walter returned to Louisiana once again, where he administered federal grants to promote and support career/technical education for 10 years.

Walter was a Board member, Treasurer and President of the Mailorder Association of Nurserymen; Board member, VP and President of the Louisiana Chamber of Commerce; charter Board member and President of the Pike County Economic Development Authority as well as being involved in numerous charitable and community endeavors. He has been a registered high school football official for more than 35 years.

In 2012, Walter ran for the office of Pike County Eastern District Commissioner, but narrowly lost. He and Cindy still reside in Louisiana.

WILLIAM PHARR STARK (1862 — 1943) / Third Generation

Treasurer, circa 1900 - 1911

WILLIAM PHARR STARK

William P. was the youngest of William W.'s three sons, a real rabble-rouser. William married Lucy Reid Harvey in 1884 and they had four children. He was educated at McCune College and joined the business right out of school, starting as office manager and working his way up to Treasurer.

In the midst of multiple disputes, he resigned from the company in 1911 and started his own nursery with his son, William H., on January 1, 1912 — while still on the Stark Bro's payroll. His business practices were retaliatory and less than legal. His brother Edgar sued him for trademark infringement, and won the case after it went to the Supreme Court. His company was renamed Neosho Nurseries in 1921, and he remained for a few years until he was ousted by the new owners. It is said he moved to Philadelphia, where he lived for decades.

William was a Scottish Rite Mason, a member of Moolah Temple and a Knight Templar. He was also a delegate to the General Assembly of the Presbyterian Church at Dallas, Texas, President of Board of Trustees of Missouri Valley College, President of Board of Education of Louisiana Public Schools and author of numerous articles on horticulture and pomology.

✣ ✣ ✣ THE THORNTON GRIMSLEY BRANCH ✣ ✣ ✣

THORNTON GRIMSLEY STARK

THORNTON GRIMSLEY STARK (1830 — 1882) / Second Generation

Thornton, though not a lifelong Stark Bro's executive, grew to be a prominent citizen of Calumet Township. He worked with his brother in the nursery business for a short time, but left and bought a basket factory instead, which employed about 20 people.

At the age of approximately 50, he developed cancer of the neck and traveled to Arkansas to attempt to treat his disease in the "healing waters" of Eureka Springs. His self-treatment failed and he passed away in 1882.

He was a devoted worshipper at the Dover Baptist Church, and so his remains were brought back and laid in the old churchyard; at some unknown point he was re-interred at the Stark Cemetery in Calumet Township.

EUGENE WASHINGTON STARK (1865 — 1909) / Third Generation

Secretary, 1900 - 1909

Born at the end of the Civil War to Thornton Grimsley Stark, Eugene grew up in the uneasy aftermath. In the town of Stark, where he was born and raised, Eugene established a general store and a post office. He later sold the store and joined his uncle and cousins at Stark Bro's Nurseries. By that time he had married Ann Withrow and they had three boys: Thomas W., Lawrence E., and Edwin J., all of whom held future executive positions at Stark Bro's. Eugene, like his great-uncle James Hart Stark, served as county Judge of Pike County, and later his political experience and his wide popularity won him election to the Missouri Senate. His work at the nursery office combined with his position as state Senator were visibly taxing, and yet friends were urging him to run for governor. The stress was too heavy a burden, and after a major operation, Eugene died at the very young age of 44.

THOMAS WITHROW STARK (1891 — 1943) / Fourth Generation

Secretary, Unknown - 1934

Thomas Withrow Stark was married to Alice True; they had no children. Until 1934, Tom served as Secretary of the company, and for many years was manager of the Collection Department. He, too, was schooled in horticulture and he traveled to various parts of the world, both learning new procedures and seeking new markets. He also served as Mayor of Louisiana, Missouri.

LAWRENCE EIDSON STARK (1893 — 1958) / Fourth Generation

Western Sales Director, Unknown - 1953

Lawrence was the second son of Eugene and Ann Stark. He joined the company in 1914. After World War I, he went to Europe as associate editor to advise on the rehabilitation of agriculture in war-torn countries. He traveled extensively in America and Europe gathering material for other articles on fruit-growing. In 1933, he married Marie Dickison, but had no children. By the 1940s he had become Secretary-Treasurer and late in his career he was appointed Western Sales Director. He died in the hospital of heart-related problems in 1958.

EDWIN JACKSON STARK (1898 — 1964) / Fourth Generation

President, 1953 - 1964 (died in office)

Edwin was the youngest son of Eugene. After graduation from high school in 1918, and subsequent Army service, he returned to attend Missouri University, where like all Starks of recent years, he studied horticulture, specializing in nursery production and methods. When Clay Hamilton Stark passed away in 1953, Edwin assumed the role of President and made a conscious choice to work closely with his brothers, Tom and Lawrence. He served as President of the American Association of Nurserymen and was active in both horticultural and civic affairs. He was president of the Missouri Fruit Experiment Station at Mountain Grove for many years, and later served as a member of the Board of Curators of the University of Missouri. Edwin faithfully served as president until his passing in 1964.

The Current Owners

CAMERON G. BROWN | President and CEO, 2001 —

Cameron grew up on his parents' hog, cattle and grain farm in Griggsville, Illinois. In 1967, he graduated from Blackburn College and married Sybil Hudson; they had three boys. Cameron was drafted into the Vietnam War and upon returning, received his Masters Degree from De Paul University in Comparative Literature.

He later attended Georgetown School of Medicine as a Graduate Fellow, after which he created several HMOs in Wisconsin. Until he bought Stark Bro's, Cameron was in the healthcare plan administration and technology field as a founder and consultant of many HMOs. He also founded Sage Technologies, Inc., a healthcare administration company in Illinois.

His ties to Stark Bro's go back many years: Cameron's father, Lou, started the Garden Center which is still in operation. More recently, his late son, Kyle, launched the company into technological competitiveness. He was tragically killed in a traffic accident in 2011. The Brown family has leased tree-growing acreage to Stark for decades, and still does. Aside from his passion for growing things, Cameron is an instrument-rated pilot, avid sportsman and outdoorsman.

TIMOTHY L. ABAIR | Chief Financial Officer / Co-Owner, 2001 —

A portion of Tim's teenage years was spent on a tree farm, and even then he had an interest in owning a tree nursery someday. Tim is a Certified Public Accountant, in public practice since 1994. As a founding co-owner of Sage Technologies in Rockford, Illinois, he has provided consulting services to specialty and primary care physicians, hospitals, insurance companies, individual practice associations, nursing homes/assisted living facilities and preferred provider networks.

As CFO of Stark Bro's, Tim meets monthly with the Executive Team and provides strategic financial analysis and advice to the company. He also works with the company's attorneys, banking partners and vendors.

5

Young James

A Risky Move To Missouri

The Budding Nursery Business

5 | THE ROOTS OF A HORTICULTURAL DYNASTY

1803-1844

Young James

James Hart Stark was the spark that ignited this company's impactful history; he was born on July 30, 1792. Without that commonplace event, there would not now be a world where 60 percent of the apples grown come from the parentage of just two: the Red Delicious and the Golden Delicious, both discovered by Stark men. There is quite a story to be told about the time in between.

Young James Hart Stark

James was just 11 when his mother Susanna died. Not long thereafter, his father took a third bride, Rachel Brice, and they had eight more children (a total of 19 for the Captain across three marriages). The sibling rivalry must have been something to behold!

As a youngster, James developed a discernible aptitude for growing things, and at 14, was a qualified orchard hand. He was a quick study with a curious mind, and picked up the art of budding and grafting fruit trees as if it were a game of hopscotch.

By the age of 18, he was experienced enough to join a government surveying party. He is said to have become so expert at this work that he could describe any piece of land, by being told where it lay, or, if given the number of a piece of land, he could state its exact location.

A map of the area as it was in 1816. In 1813, the decisive Battle of the Thames subdued the British and Indian threats in the North and West.

James enlisted in the military as a teenager and like his eminent kinsman,

General John Stark, got on his horse and rode out to fight the enemy in the War of 1812. He admirably served under General William H. Harrison at Tippecanoe, and was in Colonel Richard Johnson's regiment at the Battle of the Thames, which ended the war in the West in 1813. The victory restored American dominance in the Northwest, making it much safer for settlers to expand westward.

And so it was then that James Hart Stark decided the time was right to move to Missouri. He was discharged on November 20, 1813 and returned to Bourbon County, Kentucky; it took him two years to find a wife named Jane Watts.

A Risky Move To Missouri

That same year, James and 19 other men set out on a long horseback trip to Missouri to look at land they might live on. One of the men rode ahead to the next creek, trying to settle his very spirited and thirsty horse. Ten minutes later, the rest of the group caught up to him, where he was sitting on a log, his horse grazing nearby.

James asked what had happened, and the man replied, "The Indians shot me," pointing to his chest. "From behind those trees. Lots of them — did you hear them shoot? Providence sent me on ahead, or all of you would have been killed."

The Indians apparently believed the shot had missed, and fearing a larger party was nearby, fled. James had escaped a nearly certain death.

The group rode on to the Mississippi River without incident, and it was there that James picked the spot for his homestead: on high ground, west of the river, about seven miles south of where the town of Louisiana now stands. Well-satisfied, James returned to Kentucky and made plans to permanently resettle in Pike County, Missouri. Apparently Jane didn't have much say in the matter.

In 1816, James Hart Stark and his young family migrated to Missouri. His saddlebags were filled with Jeniton apple scions, the beginning of the first nursery west of the Mississippi.

Excited but filled with trepidation, Jane bundled up their six-month-old son Henry and joined James, her parents and an assortment of other settlers on the journey to Missouri.

It was near sundown on October 10th, 1816, when the band of 20 pioneers, bone-weary after days of travel in wagons and on horseback, crossed the Mississippi River at a point where the town of Louisiana is now situated. They hastily made camp at the spot James had chosen a year before. On this trip, James had stuffed his saddlebags with bundles of Jeniton apple scions from his father's Kentucky orchard.

Little did he imagine how important those scions would become — not just to him, but to future generations of the Stark family, and to the whole world of horticulture. When grafted to wild crab apple trees, those scions would grow some of the first cultivated fruit west of the Mississippi, and eventually grow into an international dynasty of forward-thinking nurserymen.

Apple scions, just like these, grew into a horticultural empire.

Soon after reaching their destination, the earlier settlers turned out to help construct homes for the new arrivals: cutting down trees, hauling and hewing the logs, and building the houses and barns.

When this was accomplished an old hunter said, "James Stark, you get your gun and we will get your winter's meat." This was a land where wild turkeys, deer and other game were abundant. Wild plums, berries and grapes were plentiful, and even the luxury of wild honey was easy to find.

Now that he had sound shelter for his family (namely the famous Stark Cabin), James began clearing land and preparing to start his orchard. He wanted no mongrel seedlings nor their sub-standard fruit; so, onto the roots of the sturdy wild crabapples he set the scions of the fine Jeniton parent trees. The artful grafting techniques he had learned from his father stood him in good stead.

In Kentucky, the locals had scoffed at his plans, saying Missouri was a place where fruit trees would never grow. They were almost right, as 1816 was the "Year Without A Summer." Volcanic ash from the Mount Tambora eruption caused lower temperatures worldwide, ruining crops and farmers on every continent. Hundreds of thousands of people died of starvation. Fortunately for James, his destiny was also planted in that ground. The trees survived.

The Budding Nursery Business

Two years later, the grafted trees were large enough to sell. By this time, James had planted 45 acres or so, and had the only commercial orchard in his section of Pike County. Only the finest apples went into the barrels to be sold, and James was the one to personally grade the fruit.

To further his fortunes, he considered planting a large additional orchard. An off-handed discussion with a neighbor was naively prophetic:

"You are foolish to be planting so many trees ... you will never live to enjoy the fruit from them," said the man. To this James replied, "If I do not, some of my children may."

Inspired by his own words, James planted yet another orchard.

Scientific methods and natural selection were crucial in producing strong, adaptable varieties, and James was fully armed with this critical knowledge. The quality and sturdiness of his stock far outshone the few susceptible trees in existence there. It was as if someone had sprinkled magic growing dust on the earth and good apple trees sprung up.

"Old Patriarch" was planted in 1818 by James Hart Stark, just a few miles from the Stark Bro's offices in Louisiana, Missouri. Without any special care, it bore fruit until it died at the very old age of 102.

Predictably, word spread quickly and people began bringing their wagons from far-flung distances to see about these marvelous new trees. Before long, Stark apple trees were helping to bring prosperity to Pike County and the surrounding areas. No money was exchanged for the trees; James only wanted to populate the countryside with good apple stock.

In the end, his generosity came face-to-face with hard reality: he simply couldn't give away hundreds of trees and expect to support his family. And so he began to charge a few "coppers" for each tree — and that was the first profit for Stark Nurseries. Soon, orchardists would come to buy the trees, and apple buyers would come from the North and East to buy the apples. Eventually, thousands of barrels would be shipped annually.

Grading apples as they were picked cut handling costs. Apples in barrels kept in cool basements or buried in the ground in winter kept remarkably well.

As the nursery grew, so did James' family. He and his wife, Jane, eventually had 17 children, more than enough hands to run a thriving nursery business — but only William, the seventh son, would carry on the enterprise's legacy in earnest.

It was not long before James had become to Missouri what Johnny Appleseed had been to the Ohio Valley, only with far better apple trees. The difference was "stark," one might say: Chapman's trees died out within a few years, while James' thrived, some for decades. Chapman had simply planted apple seed indiscriminately, resulting in low-quality seedlings that were not true to the original tree. James planted orchards practicing careful selection and grafting. The result was a booming nursery business.

Said the *Centennial History of Pike County*: "These apples ... supply the wants of the pioneers, tickle the literary palates of the aesthetic Bostonians and supply the table of a portion of the royalty of Europe."

James Stark was certainly an accomplished nurseryman, but he also had a rather high opinion of himself in matters of law and order. He felt a calling to make a foray into politics, which became something of an obsession.

When he ran for Pike County General Assembly as a Whig in 1836, his campaign focused on his insistence that Missouri banks assure a sound currency; that it adopt a planned system of public roads and railroads, and pay the owners a fair sum for any private lands taken; and he strongly urged improvements to the education system. To his great dismay, he was not elected.

Four years later, James Hart Stark ran for and was elected to a four-year term as Judge of Pike County, representing Buffalo Township. This time, he prevailed. Even though he held office for just one term, "Judge" was a title he used for the rest of his life. He was darned proud of it.

James' pursuit of his judgeship would leave a void elsewhere, however. Managing the nursery would now be up to William, who was a mere 14 years old.

6

A Peaceable Kingdom

The Woes of War

A Most Fruitful Period in Time

6 | THE WAR AND THE BIG BOOM

1844-1872

A Peaceable Kingdom

There was peace in the valley, as the saying goes, in the decade after Judge Stark retired from office. An interesting picture of life at the Stark nursery in the years just preceding the Civil War is vividly painted in a letter written in 1927 by the late Ambrose Stark Carroll (grandson of James) to his cousin, Paul Stark Sr., in Louisiana, Missouri.

> *"As I remember it was in the spring of 1857 that I went with our old farm man to my grandfather's farm south of Louisiana where the nursery was first started. We went to get trees for ourselves and our neighbors.*
>
> *"As the mud was deep and the road very bad, we drove two yoke of oxen to a very high-wheeled, scooped-bed wagon, but most of the hauling in those days was in ox carts with only two wheels. When we arrived at Grandpa's, there were a great many people there from the surrounding country, some coming from as far as Boonville over on the Missouri River.*
>
> *"All stayed all night or for dinner and helped dig the trees. I don't recall that the trees were labeled, so each one had to remember the variety he got. I remember such names as Jeniton, Bellflower, Vandiver, (both yellow and red), Winesap, Rambo, Pryor's Red, Red June, Spitzenburg, Early Harvest, Golden Russet, and a little red apple they called a Milan. Oh boy, I can taste them yet!*
>
> *"Aside from the hopes of the 'blessed hereafter', nothing gives me more joy than the memory of those visits to Grandma's. This was the beginning of what has taken more than 100 years to develop into the great Stark Nurseries, the largest in the world, and added to the vision of its founder, James Stark, has been the energy, ability and constant effort of his son, grandsons and great grandsons to place it in advance of any other in the entire world."*

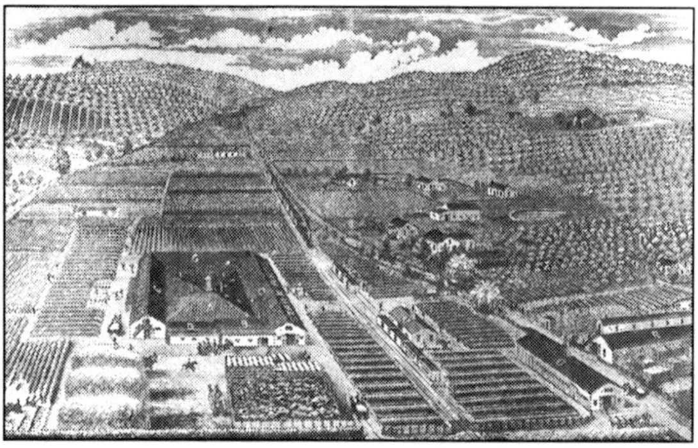

An artist's depiction of the railroad running through the Stark nursery grounds

Life was good then. William undertook a huge expansion of the company, and Stark Bro's was growing acre upon acre of fruit trees.

The Illinois Central Railroad was critical in allowing Stark Bro's to sell the enormous number of trees they were producing. Soon after the railroad was completed to St. Louis in 1855, local farmers (and a few opportunistic investors) quickly realized that produce could get to the Chicago and St. Louis markets in less than 24 hours by train. Within 15 years, the railroads would be the company's mainstay shipping method.

William, secure in the knowledge that the business was flourishing, chose Cynthia Eliza Pharr to be his wife in 1853. They would eventually have five children together. His brother, Thornton Grimsley Stark, came on board in 1855 to help run the organization for several years. William never considered anyone else; the Starks were loyalists.

William foresaw the dangers of letting the business get too big, too fast, and wanted extra hands on board to keep things under control. As smart as they were to be so anticipatory, neither of them foresaw that in a few short years, the country would be decimated by a war within itself.

In the meantime, William, who often consulted with farmers about things horticultural, soon realized that other nurserymen in the area could benefit from forming an association of sorts. They would be able to share information and stand up against the growing stack of government regulations that was beginning to hinder some aspects of doing business.

An annual meeting of William's Missouri State Historical Society in 1907

And so in 1859, William and a close friend organized the very first Missouri State Horticultural

Society meeting, after which William served as Vice-President for 12 years. His annual reports were full of valuable wisdom, a treasure-box of accumulated skills and insight from the generations before him. The advice he gave was a veritable how-to for orchardists everywhere. William was nothing less than a horticultural luminary.

The Woes of War

The war came in April of 1861, as everyone dreaded it would. Because of a weak heart, William W. Stark was unable to fight. He did provide service to Pike County, however; when the sheriff was killed by guerrillas, William was appointed to replace him as deputy sheriff, a job he held throughout most of the war.

The fortified home of William Watts Stark

Just prior to the beginning of the conflict, William was occupied with building a new home for his young family. To insure against the same fate that befell his predecessor, he built a doorless, windowless room into the center of the house. The only means of entrance or exit was a trap door, which led to the basement. There, he would be protected at night from the rebels.

William H. Stark, a New Hampshire cousin who fought in the Civil War

The nursery still needed managing, a stressful undertaking considering the addition of his sheriff's duties. But at 34, William had already been running the operation for 20 years and all concerned were confident that it was in very capable hands.

Despite his experience and expertise, the nursery business fell off drastically during the war, but with some luck and providence, he managed to pull through. In the face of so much destruction, the Stark Bro's nursery was largely spared, at least physically speaking.

With the war behind it, and a peaceful, prosperous future ahead, the town of Louisiana grew. Judge James Stark moved from his famous log cabin and built a home on the edge of the new town. He became an influential citizen, active in local affairs, and also took another swing at politics.

An ardent Whig, he was a supporter of Henry Clay. In 1864, at a meeting in Bowling Green, Missouri, the Clay Club was organized. James Stark was named to draft a constitution and a resolution advocating the election of Henry Clay to the Presidency of the United States, which ultimately failed. Political life was just not in the stars for James Hart Stark.

A Most Fruitful Period In Time

After the war, William bought out his brother Thornton, and returned to the task of growing the business. The pickings were ripe, one might say.

On June 15, 1867, an item in the *Centralia Sentinel* noted, "Immense quantities of blackberries are being shipped from here daily. The shipment of peaches, pears, and plums have commenced and our city will soon put on its busy fruit scene, peach boxes here, peach boxes there, and peach boxes everywhere." It seemed the area's advantages for fruit-growing were becoming obvious to everyone.

Stark Bro's orchards at the time of William Watts Stark

The locale also had a distinct shipping advantage. Fruit from the south had to be picked either the day before or very early in the morning, when dew was still on it — giving it the tendency to rot. Then the fruit had to suffer 24 hours of a hot, dusty train ride, be handled again upon arrival and then taken to market. Such treatment of delicate fruit like peaches and strawberries was disastrous.

In Louisiana, Missouri, the fruit could be picked later in the morning, put on railroad cars practically fresh from the tree, transported north overnight when the weather was cool and arrive virtually unscathed. That made fruit from Louisiana worth about 25 percent more than fruit shipped up from the south.

Growers carting their fruit to the railyard

Over the summer, the train depots were bustling with railroad cars full of seasonal fruit — but problems were creating a less-than-peachy scenario.

That fall, the Southern Illinois Horticulture Society meeting became a free-for-all against the railroad and the platform workers. Railroad cars loaded with fruit had been left to sit for an entire day with no ventilation; other growers were angry over the careless handling of their tender crops. It was even whispered that the workers were purposefully dropping crates so they could take home the spoils. William W. Stark was among the chief complainants.

By 1901, Stark had its own railcars

The local paper sided with the growers, demanding an on-site overseer "with a cat-'o-nine tails, and discretion and strength to wield it to whip these vandals into a decent respect for the property rights of shippers."

By the late 1860s, as many as 20 carloads of peaches a day were arriving in Chicago. The following excerpt from an article from the *St. Louis Republican* describes the area's fruit industry at the time:

> *"... from north to south, and from east to west, peaches, apples, pears, strawberries, raspberries ... almost every other kind of fruit meet you at every farm home, city, village or hamlet or by the roadside in quantities that astonish any but 'suckers.' ... In a word, if you desire to locate in the Garden of America ... then I say, pull up stakes, sell out and come here with bag and baggage."*

Soon bankers, doctors, merchants, teachers — all were investing their money in orchards and strawberry patches in hopes of reaping a quick return. As green as they were, agriculturally-speaking, the ersatz nurserymen were eager to learn.

Greenhorns in any profession are like the dinner bell for swindlers. William did not take kindly to them, as he was a supporter of new growers. In a report made to the Missouri State Horticultural Society in 1868, William Stark wrote about "the operations of another pest, the oily-tongued and dishonest tree peddler," who took advantage of novices and sold them trees that would never produce. William's mission to run the "bad apples" out of the business continued throughout his career. He outlined other obstacles new growers faced:

> *"Those who contemplated planting orchards for commercial purposes were sagely told by croakers and many well-meaning but short-sighted people that they would have no market for their productions, which often deterred the beginner. "*

Until 1869, the town of Louisiana had depended on the river for long-distance transportation; they had the Southern Illinois railroad to get their trees and crops over the river. But Judge Stark and a few other of Louisiana's leading citizens had determined a while back that a transcontinental railroad line would be most beneficial to the town (and to Stark Bro's, of course). The judge's influence won out, and so it was built.

A 30-car train, full of Stark fruit trees

When the line was completed in 1870 through Noix Creek Valley, William moved the nursery from Buffalo Township to its present location, directly across the way from the Garden Center. He would work there for another 10 years.

By 1876 there was a line from Hannibal to Louisiana; this was later extended to Clarksville and finally to St. Peters, where it connected with lines into St. Louis and opened new markets galore for Stark fruit trees. The ability to ship by rail was a boon. Stark Bro's had now gone nationwide.

—1844-1872—

The interior of the Stark Cabin, when it was open for tours

Golden Delicious being sold at a NY railroad station, circa 1916

A railroad car loaded with Starking® Delicious apples, 1924

Stark employees at Gov. Lloyd Crow Stark's mansion in the late 1930s

7

The Panic of 1873

Three Forces With Which To Reckon

The Advent of Advertising

Clarence in Colorado

The Brothers Finally Incorporate

7 | THE PANIC AND THE LESSON LEARNED

1873-1892

The Panic of 1873

By the early 1870s, William sat atop an admirable enterprise, at least for a short while. The financial Panic of 1873 and that year's harsh winter took a severe toll on the business, and William was never the same after that. It was at this time that his three sons — Clarence McDowell Stark, Edgar Winfred Stark and William Pharr Stark — joined the business as naïve young men with little agricultural knowledge. Nonetheless, they assumed the operations of the business upon their father's death in 1880.

Failure, which had seemed impossible when the fruit boom was on, became a regular occurrence. On December 21, 1872, the thermometer dropped to -27 degrees F. and remained below zero for an entire week. It was the coldest period ever recorded for the area and resulted in the failure of the peach crop and the death of many trees and grapevines. Pears, apples, cherries and strawberries produced at least a meager crop, but it was an expensive one. A quart of strawberries was sold for 60 cents, a frightening amount of money for a farmhand making a dollar a day. The apples were also damaged, even though they had avoided a late-spring frost. To make matters more miserable, what the cold had not killed, the curculio (weevil) did, and the apple crop pretty much went the way of the peach crop.

A political cartoon depicting the Panic of 1873.

When the Panic of 1873 took full hold, William Stark owned five orchard farms and a flourishing nursery business. He might have survived the downturn in fair shape had he not been, as his grandson Gov. Lloyd C. Stark later put it, "too good a friend to too many." To help his financially embarrassed friends and neighbors, he had signed their bank notes. As a result of his generosity, the Starks lost everything but one nursery farm.

With grace and valiance, William's long-disabled wife Cynthia and their five children kept the family alive by selling Stark Bro's produce to the townspeople. William struggled to preserve the business. It was quite a comedown after so much success. A local barber observed that William "wore a sulking face on every visit" and seemed to have lost "his jovial way."

At the nadir of William's despair over the business losses, yet another blow: his father, mentor, confidant and cheerleader passed away. Judge James Hart Stark died May 27, 1873, at the age of 81. After a long day, James retired earlier than usual and was found dead in his bed the next morning.

Excerpted from a long article about Judge Stark in the *Centennial History of Pike County*, published three years after his death:

> *"We learn that the county early became known for its very superior apples of almost all known varieties, making the township famous even among the growers and dealers of the Northern and Northwest United States."*

It was quite a tribute to someone who came to Missouri with only a few apple scions in a saddlebag, hoping to start an orchard.

By 1880, with the panic just a bad memory and his debts paid, William was left with only the nursery farm. He summoned what will he had left, and went to work to try to rebuild the business.

The climb was so steep and so painfully sad without his father's steady presence; the task was beyond William's ability to cope. One day in 1880, he was found in his beloved orchard, where he had died of a heart attack while working among the trees.

THREE FORCES WITH WHICH TO RECKON

William's three sons — Clarence, Edgar and William P. — were left to pick up the pieces and virtually start all over again. Clarence became president of the company, and Edgar vice-president. The younger brother, William P., who was still in school, later joined the company and became office manager.

With stern determination, the brothers set out to make the Stark nurseries the greatest in the world. They never forgot the debt they owed their father, and in one of their first catalogs there appeared a picture of William Watts Stark, with the following tribute:

—1873-1892—

"To the good judgment and wisdom of our honored and lamented father, the late Wm. Stark, more than to all others is due the success of these nurseries. The broad and sure foundation long ago laid, was chiefly of his building, he having succeeded his father while yet a mere boy of 14 years. For more than 40 years he successfully labored to advance horticulture in this, his native state. His cherished wish was to place his sons in the business he had always loved; we are glad his wish that we should be nurserymen has been fulfilled."

William's Three Sons

Clarence McDowell Stark

Edgar Winfred Stark

William Pharr Stark

The sons focused on the core Stark principle of "better and better fruits, and increased production per acre." They always believed that their customers must be given every opportunity to succeed, which in turn would make them successful. This tenet remains fundamental to the company's longevity and customer loyalty.

The third-generation trio of Clarence, Edgar, and William P. proved to be a forceful one. Each of them was well-grounded in horticulture, as they had learned the nursery business from their father and grandfather.

But each had his own special talents: Edgar was in charge of production, shipping, real estate, and finance; William P. was office manager and worked with sales. Clarence, president of the company, might be said to have been far ahead of his time with his genius for advertising and promotion of the marvelous new Stark fruits that were to come.

Beginning with Clarence and Edgar in the late 1800s, it was a policy to maintain a balance of power in the branches of the family. If a descendant of one became president of the company, a descendant of the other was named Chairman of the Board; others would be VP, Secretary-Treasurer, and so forth.

The earliest interior photo of the Stark offices, taken in 1889

Clarence and Edgar made an excellent team because each had his peculiar talents, which, when put together, spelled success. This pattern prevailed for decades, with up-and-coming Starks freely allowed to gravitate toward their own proclivities. Those who leaned toward business and sales went into that department; those who had inherited the green thumb of their forebears went into nursery production and research development.

Young Starks were not pushed to work at the nursery unless they felt it was their calling, but most of them found a place anyway because there was such a diversity of jobs to be had.

The Advent of Advertising

By the late 1800s, Clarence had earned a reputation as a real creative force in the company, and seemed to latch on to the notion of "branding" before it was ever known as something other than a cattle marking. He wanted to find a way to visually reinforce the notion that Stark trees were good fruit-bearers. Linear thinker that he was, the answer was quite literal: an illustration of a ... bear.

The Stark Bear trademark was featured for over 100 years in advertising, catalogs, on barns and at fruit shows all over America.

The "Stark Trees Bear Fruit" logo — then called a "trade slogan" — was not officially trademarked until 1916 (two years after Clarence's death). The nation was made conscious of Stark Trees because of it; the logo became as much a part of the landscape as Burma Shave and Coca-Cola® signs.

The famous "boy in the tree" ad

The Starks' first national ad appeared in *Harper's Magazine* in 1887. The first time a photograph was used in a catalog was 1894; their first color catalog was printed two years later.

Clarence liked to slip in portraits of Shakespeare and a few of his lofty quotes for his own amusement; what was not amusing was that the extra pages made the catalog frightfully expensive to produce, killing a large percentage of the profit.

The brothers began placing large advertisements in any number of national newspapers and magazines. One of the most famous was a picture of a small boy perched in an apple tree with a man on the ground below, ready to throw a rock at him. The caption: "It takes stones and bulldogs to drive boys out of STARK trees."

Clarence in Colorado

William P. and Edgar had taken notice that Clarence "wasn't himself" for a time, and they urged him to see a doctor. In 1887 Clarence, said to be a victim of tuberculosis, was advised to go to Denver, Colorado to "rest." It is not known if the diagnosis was that of the doctor, or the one that the brothers chose to put out to the public at the time.

In any event, Clarence went to Colorado. Not at all content to sit and "rest," he promptly acquired a parcel of 100 acres and planted 12,000 fruit trees. It was one

Clarence (left) in his Denver orchard, circa 1891

of the few failures in trio's tenure: wells were drilled and pipe was laid through the whole orchard, but the wells dried up shortly thereafter. It was a bankrupt investment and caused some consternation between the sons and their bankers. Somehow, in 1896 (through "hard scheming," as Edgar called it) they were able to get water to the orchard and save it.

Clarence's Colorado experiments opened the eyes of fruit growers and scientists to the possibilities of dry farming in arid western areas. He showed that constant cultivation of the right kind formed a mulch which would conserve moisture. In 1891, the USDA sent an expert to Colorado to look at this orchard. His report, in the 1892 USDA Year Book, declared this lesson was "worth millions of dollars.""

The Brothers Finally Incorporate

Back in Illinois, the gigantic fruit bubble eventually burst. Too many growers had gotten into the business. Add to that the development of new varieties, with better shipping qualities and faster, refrigerated rail cars. The result: Louisiana, Missouri berries were getting to Chicago long before southern Illinois berries were even ripe. Ironically, the railroad — which had hatched the fruit industry in southern Illinois — was now killing it with its improved service.

Edgar Stark studied the failures in Illinois, reading them as a cautionary tale. He was a man who tried to provide for any contingency and anticipated pitfalls wherever he could. The generations who followed him knew that Edgar expected mistakes, but what he didn't tolerate was one that was twice-made. It would be a lesson that sons Lloyd Sr. and Paul Sr. would take to heart.

And so the failed berry growers turned their attention to apples, a territory to which the Stark brothers had already laid claim. They had been in the apple business for 73 years now, and had a decided experiential advantage over any new competitors. They had done staggering volumes of business, but almost unthinkably, had never incorporated their enterprise!

The three sons, all administratively capable, finally woke up to the fact that the assets of the business needed protecting. And so finally, in 1889, the Stark nurseries became tark Bro's Company. It cost $50,000, and had the names of Edgar and Clarence on the legal documents.

Shortly thereafter, Clarence gave power of attorney to William P. for reasons unknown. Perhaps Clarence knew something his brothers did not.

Paul Stark III (left) and Harry Guengerich

From left: Ernie Elder (Shop Foreman), Leslie "Hobo" Hickmann (Branch Mgr., Fayetteville, AR) and Paul Stark Jr.

American Association of Nurserymen gathering at Stark Bro's, June 11, 1897

Roberta Dolbeare, surrounded by "high-tech" computers

Floyd Maples (Branch Mgr., Marionville, MO)

8

The Panic of 1893

A Delicious Discovery

The Changing Shape of the Apple Market

The Delicious Monument

8 | 1893: A DEFINING YEAR

1893-1894

The Panic of 1893

Twenty years after the panic of 1873 came yet another crash. Growing piles of unpaid bills pushed Stark Bro's to the edge of the bankruptcy cliff. Somebody needed to find a rabbit they could pull out of a hat, and quickly.

An undulation of economic waves plagues every business, and Stark Bro's was no different — except for the fact that no matter what happened, or how bad it was, some serendipitous miracle always seemed to pop up and calm the rough seas. It was about to happen again.

Edwin Harpole Sr., whose son would become a Stark Bro's branch manager, went to the brothers' office in that dark hour with a rabbit, and a hat.

Ed tossed a bundle of cash on the desk. "I hear you need money," he said. "Here's $2,000, all I have, but you're welcome to it."

Ed Harpole's son, Ed Jr., who was with Stark Bro's for 52 years

And that was just the beginning.

Like something out of "It's a Wonderful Life," Fieldon Estes, another patron, had heard of their distress. He paid the brothers a visit.

"Your father and mother were my dear friends," he said. "Pay me back when you can," and firmly placed $10,000 in gold in Edgar's hands.

Workers chipped in to help, as well. They accepted script for pay, which some of the local merchants accepted until it could be redeemed by the company. In the meantime, the young Stark sons humbly wore patched pants and never complained.

Clarence realized that all of these unpaid bills and other liabilities were leaving the company with serious legal exposure. Their 1889 incorporation was flimsy, so in 1893, the firm was re-incorporated for $300,000 under the official name it has today: Stark Bro's Nurseries & Orchards Co.

The Starks were instrumental in establishing the great fruit industry in the Northwest. These Great Northern railcars were full of Stark Bro's trees.

Thanks to Ed Harpole Sr. and Fieldon Estes, the company had been rescued from the brink.

Yet another bit of serendipity came in the form of a rail line. The Great Northern Railroad reached the Northwest in 1893, opening up enormous markets at a most welcome time.

James J. Hill, the railroad's president, asked the Stark brothers to give him their opinion on the value and suitability of land along the Columbia River as fruit regions. The brothers sent horticulturists to the area and the report was very favorable.

Soon, the Wenatchee, Yakima, Hood River and other valleys of the Northwest were among the leading fruit-producing areas of the country. Within a decade, trainload after trainload of Stark trees went to Washington, Oregon, Idaho and other Western states. With further development of the railroads to the West, Stark trees traveled to New Mexico, Utah and California. Later, Stark trees made it to Mexico to the South, Canada to the North and other foreign countries.

The rail line to the northwest was a boon, and most certainly helped the company get its footing again. But Clarence realized that they needed to boost sales and bank some of that money to better cover potential down-times. He never wanted to stare bankruptcy in the face again, so he ramped up their advertising schedule and started a series of very innovative promotions. One of those ideas would open the door to discovering most historic apple since the Garden of Eden.

A Delicious Discovery

What is probably the world's most familiar apple — Red Delicious — was discovered pretty much by chance, a happy accident that vaulted Stark Bro's into the annals of fruit-growing lore.

Back in 1893, the best-known and the most widely-planted apple was Ben Davis. It was a hardy apple that stood up to tough winters, would grow almost anywhere and did not bruise easily. It had a rather significant drawback: it didn't taste very good.

—1893-1894—

A trade card for the 'King David' apple tree

Clarence Stark knew well the shortcomings of Ben Davis apples. Unable to discourage farmers from raising them, he decided the only alternative was to find a better apple himself.

And so intrepid Clarence literally got on his horse, and traveled through the Ozarks, studying varieties all along the way. He returned with several good ones — 'Senator,' 'King David' and 'Champion' — but the elusive flavor he sought was not among them.

Never easily satisfied, Clarence continued his search for a more perfect apple. But how to find it? He couldn't keep traveling all over the country on horseback, and jet airliners hadn't been invented yet.

Then the idea struck him: have growers send in samples, and then conduct a contest to determine the best variety! Not only would new apples come to Clarence, but he could also draw attention to Stark Bro's with ads for the contest.

And so the very first International New Fruit Show was held, with great fanfare, in 1893. Professor Henry E. Van Deman, then the official USDA Chief Pomologist, was brought to Louisiana from Washington, D. C. to be a judge. Invitations were sent to apple growers all over the country, asking them to send in their varieties for the contest. Hundreds of entries were received from 27 states.

One of those growers was Jesse Hiatt, a good Quaker man. Some years ago on his Peru, Iowa farm, he discovered a chance seedling in his orchard of 'Yellow Bellflower' trees. He chopped it down, but the next season, it sprang back up. He chopped it down a second time. It came back again.

"If thee must grow," he told the stubborn sprout, "thee may," goes the story.

Some years later, Hiatt's tree bore its first fruit. The apple was an elongated globe with red-and-gold striped skin, crisp flesh and a five-pointed calyx. He named it 'Hawkeye,' in the spirit of his home state, and offered it around for sale — but there were no takers. It was turned down by at least eight nurserymen because of its strange oblong shape and bumpy bottom end. "Just too odd-looking," they said. What was not odd was the flavor, which was just plain wonderful.

Hiatt received a gracious invitation from Stark Bro's Nurseries & Orchards Co. to send them samples for their novel New Fruit Fair. He enthusiastically packed up some of his Hawkeyes and shipped them off — along with some high hopes — to Missouri. (The fruit actually landed on William P. Stark's desk, and it is he who is credited by his brother Edgar for "discovering" the apple.)

On the day of the contest, five apples of each variety were put on plates and grandly displayed on long tables. The entry card was placed near each plate to

The International New Fruit Fair, held at old newspaper office in Louisiana, Missouri. Mark Twain served as an apprentice newspaper printer here.

identify the grower. It was quite a sight, all of those red, shiny apples. A seven-year-old boy was in attendance, marveling over the myriad of varieties. That little boy grew up to be Lloyd Crow Stark, the long-time chairman of Stark Bro's, and Governor of Missouri.

There was much animated discussion, some arguing and a lot of apple-tasting until the winner was finally announced: first prize went to the apples with finely-colored red and gold stripes, long, distinctive shape and five small bumps on the calyx end. A good-looking apple to be sure, but it won for its exceptional flavor.

The moment Clarence bit into one, he excitedly exclaimed, "My! This apple is delicious!" He paused, thought, and spoke again. "That will be its name! Who sent them?" The judges overwhelming declared the strange-but-delicious apple the winner.

But who had sent the apple? Chatter rose as the crowd speculated. To Clarence's great dismay, it became clear that no one knew who sent the prize-winning beauties — the entry card was missing. Either the grower had failed to send it, or it had been carelessly lost. A frustrating, frantic search failed to bring it to light.

Mr. and Mrs. Jesse Hiatt, with the original Delicious tree

Meanwhile, back in Peru, Iowa, Jesse Hiatt was busy with his orchard. He never heard from the folks at the New Fruit Fair, and figured they must not have liked the shape of the Hawkeye, either. As a faithful Quaker, however, he refused to give up, and persisted in trying to get a taker for his unique apples.

Clarence knew he had to find the grower of this marvelous fruit. Its flavor was superior to anything he had ever encountered, and he despaired that it might be lost forever.

—1893-1894—

He immediately sent letters to every grower who had entered the 1893 contest, urging them to send their best apples to the next Stark Bro's New Fruit Fair.

Hiatt received that second invitation, and dutifully sent his long-and-bumpy apples to the Stark folks, along with a letter suggesting that they buy his tree. Within five minutes of the fruit's arrival, the brothers knew they had been blessed again. The contest went on, but this time, there was little discussion and no arguing whatsoever.

The winner of the 1894 International New Fruit Show: Hawkeye, grown by Jesse Hiatt of Peru, Iowa. There was an entry card, and with it, an enormous sigh of relief. The Delicious apple was saved for the world to enjoy.

Clarence was elated. He could not get to Iowa fast enough. The Stark brothers made the trip immediately and bought the sole rights from Hiatt, named the apple "Stark® Delicious" (only after naming 'Golden Delicious' in 1914 did it become *Red* Delicious) and secured a registered trademark. They could not wait to introduce their find to the public.

The Changing Shape of the Apple Market

Over the next 20 years, the Starks spent nearly $750,000 to market Red Delicious trees. They dispatched a traveling sales force across the country and placed dozens of print advertisements. The apple was even on exhibition at the 1904 World's Fair in St. Louis.

1916 centennial catalog cover touting the 'Delicious' apple tree.

At first, the big apple buyers across the river in Chicago were tepid toward Delicious as a commercial fruit because of its unconventional shape, but they did agree on one thing: it tasted mighty good. They gave it their blessing for a home orchard, but passed on it themselves "because it wasn't a round apple."

Such a patronizing response might have discouraged a less driven man, but not Clarence. He was set on proving to the world that this apple was the future of appledom.

So confident was he that he pronounced, "let the apple prove its own case," and set his ingenious plan in motion: in each large tree order the nursery shipped out, a bundle of ten trees marked "Key Z26 - for experimental test only" were included at no charge. Clarence also sent scions and trees to fruit-experiment stations and leading pomologists — the equivalent of "immersion marketing" today.

"If this apple is the winner I think it is, we will hear plenty about it," said Clarence.

Roughly four years later, when the "test trees" began bearing fruit, there was a flood of enthusiastic letters from growers who wanted to know a lot more about this mysterious Z26 apple. Delicious had indeed "proven its own case," and Clarence was, for once, a satisfied man.

Four-year-old 'Delicious' tree, loaded with fruit, caused a sensation

After the completion of the Great Northern Railway, Clarence sent boxcars of Delicious trees to new orchards in Washington state. The new apple caught on there and across the country, quickly becoming a favorite of both growers and customers because of its, well, deliciousness.

Luther Burbank, known as "The Wizard of Horticulture" for his plant-breeding marvels, called the variety "the finest-flavored apple in all the world." European orchardists heard about Delicious and ordered it to replace their trees.

The bumpy red fruit proved to be the most profitable apple ever discovered, and became the favorite of America and the rest of the planet. Through the years, the Starks have sold more than 10,000,000 of the original Delicious trees.

As for the fate of Jesse Hiatt's original tree, it became quite a celebrity. People literally wanted a piece of it. To protect their investment, the Stark brothers erected a cage around the tree to prevent anyone from snipping off branches to graft and make into new trees of their own.

A monument was erected in Winterset, Iowa (Hiatt's birthplace) in 1922. At the dedication, horticultural authorities from all corners of the country came to pay homage to this grand old tree, known as "The Mother of Millions."

Stories of the demise of the tree vary, but by most accounts it succumbed during a bitter winter blizzard in 1940, which killed most of the fruit trees in western Iowa. A descendant of the original Delicious tree still stands in the field just north of Peru, Iowa and residents enthusiastically celebrate it each year with an Apple Days Festival.

The inscription on the boulder reads:

TO
COMMEMORATE
THE DISCOVERY IN
MADISON COUNTY IOWA
OF A VARIETY OF APPLE BY
JESSE HIATT A. D. 1872
AND
CALLED BY HIM
THE HAWKEYE
SOLE RIGHT TO PROPAGATE
ACQUIRED BY
C. M. STARK A. D. 1894
AND BY HIM RENAMED
INTRODUCED AND DISSEMINATED
THROUGHOUT THE APPLE WORLD AS

THE DELICIOUS APPLE

ERECTED A. D. 1922
IOWA STATE HORTICULTURAL SOCIETY
MADISON COUNTY HISTORICAL SOCIETY
HISTORICAL DEPARTMENT OF IOWA.

9

Luther's Young Life

The Seed of the Stark-Burbank Partnership

The Passing of the Baton

9 | THE WIZARD OF HORTICULTURE

1894-1926

As if discovering the Delicious apple weren't enough of a blessing, it was also in 1893 that the Starks began their storied cooperation with Luther Burbank and his fantastic new varieties of fruits.

Luther was an American botanist, horticulturist and pioneer in fruit and plant science. Over the course of his 55-year career, he developed more than 800 strains and varieties of plants, including fruits, flowers, grains, grasses and vegetables.

Luther Burbank, "The Wizard of Horticulture." This photo was featured in an 1896 Stark catalog.

Dr. W. L. Howard, Director of University of California Experiment Station, wrote of Luther Burbank:

> *"... during his working lifetime (1873-1926), he probably contributed or introduced more plants than any other single American in our history."*

There are not many who would argue with that statement, even in 2016.

Luther's Young Life

Luther was born in Lancaster, Massachusetts in 1849, was a man of modest means and little formal education. As the son of a farmer, he attended high school only in the winter and worked the fields the rest of the year. His father, Samuel Walton Burbank, was related to the Walmart founders; his mother was Olive Burpee Ross, from the famous Burpee seed family. Success was in his genes.

The studious young man discovered an interest in design at school, and so his father got him a job as a woodworker in a Worcester factory. Breathing sawdust all day took a terrible toll on his health, which forced him to quit that job and rather randomly leap into the field of medicine.

Less than a year into his studies, Burbank's father died, obligating Luther to return to the familiar work of growing plants.

A constant reader and profound thinker, he knew that the success of the farm depended on more knowledge than he had. He dove into a program of intense scientific self-training at the local library, focusing on one book: Charles Darwin's *Variation of Animals and Plants under Domestication*. That work inspired Burbank to start budding and grafting. Using the pittance he inherited from his father's passing, the 21-year-old bought a land parcel near Lunenberg, Massachusetts in 1871.

Built in 1890, Luther's greenhouse survived the 1906 San Francisco earthquake

Luther Burbank inspecting his new Burbank Shasta Daisy hybrids

One of his first plant-breeding successes was the Russet Burbank potato, which was introduced in Ireland to fight the potato-blight epidemic. It is still the world's most widely grown potato. (Luther Burbank is also responsible for the Shasta daisy, the freestone July Elberta peach, spineless cactus, Santa Rosa plum and hundreds of others.) Among his more interesting trees and plants was a test cherry tree in which Luther grafted 200 different varieties, so that each variety could be tested under identical conditions.

In 1875, he sold the rights to the Burbank potato for $150, which he used to travel to Santa Rosa, California. There, he established a nursery, greenhouse and the famed Gold Ridge experimental farm, and began plant-breeding in earnest. At any given time, Burbank might have had 3,000 simultaneous experiments going on.

The Seed of the Stark-Burbank Partnership

Stories of Luther's acumen began spreading around the world, and his name became synonymous with better plants. The public began to understand what "propagation" meant, and were eager to support what came of it.

—1894-1926—

King Alfred of Belgium and Professor Hugo deVries, the noted Holland botanist, were just two of Burbank's world-famous friends, a group which came to include kings, queens, presidents, writers and world leaders in science, art, government and industry.

Luther, center, with famous friends Henry Ford (right) and Thomas Edison (left)

Burbank, who was to become known to every schoolchild as "The Wizard of Horticulture," published a stunning catalog called *New Creations* in 1893. He had not yet reached the height of his fame, but already some were hailing him as a genius. Others, most likely jealous of his prowess, questioned his methodologies and were skeptical of his results. Still others were shocked that he "dared to interfere with the handiwork of God."

Among those with the foresight to recognize the possibilities of Burbank's work was Clarence Stark, who went to California and sought Burbank out. After talking to him in Santa Rosa and seeing the results of his experiments, Clarence was convinced that Burbank was right and his professorial critics were wrong.

Luther, taken during Clarence's visit

To Clarence's great dismay, he saw that Luther Burbank was operating a small seed and nursery business in an attempt to finance his experiments and provide himself a living. It was clear that he would never be able to realize his potential under these meager circumstances.

Clarence said to Burbank:

"I don't think you will ever make a real success in the nursery business because your heart is not in it. But if you will carry forward the type of hybridizing you are doing, I think you will go very far in your chosen field. To demonstrate our sincere belief in your work, our company will give you $9,000 if you will let me pick three of these new fruits you have shown me.'"

Burbank was now convinced that the Starks should focus on the business side, and he should devote himself to the plant breeding. This relationship would cement the destiny of Stark Bro's, and put it in the league of "agricultural empire."

Said Burbank, some years later:

"I had been getting very nominal and insufficient prices for my new fruits, and when Mr. [Clarence] Stark made this offer to me, that is when I really became the Luther Burbank that horticulture has known. I could see a vision of the great possibilities of plant improvement if only sufficient time, money and effort were put behind the work ... Over the years, my relations with the Starks have been most pleasant and rewarding."

Burbank's July Elberta freestone peach

Burbank often credited the Stark family with making his work profitable. In return, he later joined with Thomas Edison to support Paul Stark Sr. in his fight to get patent legislation passed for plant breeders.

Along with Clarence's $9,000 worth of help, Luther also had something of a fan club — The Luther Burbank Society. The group took it upon themselves to publish his discoveries and manage his business affairs, affording him some additional means by which to live.

In 1903 he was awarded a $10,000 grant from the Carnegie Foundation, and in 1912 Congress appropriated 7,680 acres on which he could continue his cactus experiments. Relieved of the burden of trying to eke out a living, Burbank went on to worldwide recognition as one of the greatest horticulturists in history, a true vanguard.

Luther Burbank's Gold Ridge Experiment Farm in 1900 (Sebastopol, California)

Luther married young Elizabeth, 10 years before his death in 1926.

The ripple effect of his work on the agricultural economy is impossible to know, but consider the plum alone. Twenty of his varieties are common throughout the U. S., and half of those are grown for shipping overseas. Almost every plum at a grocery store is descended from one of Luther's.

In 1916 he married his much younger secretary, Elizabeth, causing a ruckus amongst the blue-bloods, but Luther was oblivious. He continued to lecture and authored more than 30 books, none more famous than the 12-volume set, *Luther Burbank: His Methods and Discoveries and Their Practical Application.* The set contains 1260 photographs, which document Burbank's extensive work in Santa Rosa and Sebastopol from 1875 to 1914.

The Passing of the Baton

Gastrointestinal complications and violent hiccups weakened Luther in the last two weeks before his death, which was ultimately caused by heart failure. At his bedside were Elizabeth and his sister when he died on April 11, 1926.

The famous botanist was buried in an unmarked grave, under a giant Cedar of Lebanon at the Luther Burbank Home and Gardens in Santa Rosa, California. The tree in the photo (right) is sadly no longer there.

Paul Stark Sr. at Luther Burbank's unmarked grave in California

As Burbank's life drew to a close, the question arose as to who would carry on his work, and naturally there were many interested in doing so. Before his death in April of 1926, Luther Burbank spoke quietly to his wife, and said:

"If anything happens to me, you will have to dispose of the business and the work, because you can't go on with it. There aren't a dozen organizations in the world that are equipped to go forward with it; of them all, there is really only one I think of that could make the most of it."

He named Stark Bro's Nurseries & Orchards Co. to carry on the work.

Considerable argument has been spent upon whether the plants were technically willed to Stark Bro's; they were not. He left everything to Elizabeth: money, personal property, real estate, dozens of municipal utility bonds — and the plants and precious seeds. Elizabeth had first approached both Stanford and Berkeley to have either or both universities take over the experimental farm, but sold to Stark when those proffers didn't materialize.

Stark executives with Burbank's "treasure chest" of seeds, trees and plants, along with his notes.

Mrs. Burbank entered into an agreement with Stark Bro's on Aug 23, 1927 to take the material they wanted from Burbank's properties. The contract included ownership of the business name and all of the customer information. A September 6, 1927 contract provided exclusive rights to sell uncompleted experiments with fruits at Sebastopol (except the Royal and Paradox) for 10 years. Stark Bro's had right of renewal. Tax receipts indicate payments of $27,000 to Mrs. Burbank.

Luther's cottage in 1978, boarded up and overgrown with vines

Exciting new kinds of fruits and flowers Burbank had developed (but never marketed) included 120 types of plums, 18 peaches, 28 apples, 500 hybrid roses, 30 cherries, 34 pears, 52 gladioli and many more. Stark Bro's subsequently introduced many of these varieties in their catalog.

The cottage in 1998, happily restored

Until 1931, the Experiment Farm fell into some disrepair, so Stark Bro's sent emissaries to retrieve the most promising fruit, nut and ornamental shrubs, and in 1931 sold the flowers, vegetables and seeds to Burpee Seed Co. J. B. Keil came from Stark Bro's to coordinate the efforts and worked there from 1931 to 1934.

Over the following years, Elizabeth worked with the Stark brothers to patent 16 Burbank fruits and flowers. The patents name Luther Burbank, deceased, as "inventor" by Elizabeth Waters Burbank, executrix of his estate.

In 1935, Stark ended the agreement with Mrs. Burbank (or *vice versa*). Mrs. Burbank then dispersed the majority of the gardens for subdivision. She sold the remaining property (excluding the house and greenhouse) to the Santa Rosa Junior College for use as a training ground. This lasted until 1954 (J. B. Keil stayed on as the caretaker).

The Stark-Burbank Test Garden in Louisiana, Missouri

Twenty years later, the City took over ownership of the property (which it retains today as a free public showplace). The gardens include a thornless rose, spineless cactus, rainbow corn, a hybrid mulberry tree (which Luther hoped would spark an American silk industry) and his spectacular red combustion plant.

At the test orchards in Louisiana, Missouri, in Sebastopol, California and Atlas, Illinois, the Starks kept faith with the Burbank trust, and still do. The Garden Center is housed in The Luther Burbank Building.

Stark Bro's has produced or discovered superior varieties or strains of more than 300 varieties of pears, plums, peaches, cherries, grapes, prunes, strawberries, raspberries and ornamentals. These discoveries are a direct result of being chosen to carry on Luther's experiments. The Starks have often gratefully remarked that the company would never have been as successful without their association with Luther Burbank.

Clay Stark Logan (left) accepts the award on behalf of Luther Burbank in 1986

The great Luther Burbank was posthumously inducted into the National Inventors Hall of Fame on February 10, 1986 by Clay Stark Logan, an enormous honor for him.

The Fine Art of Propagation

Luther Burbank was probably the best grafter/budder who ever was. Propagating trees (asexual reproduction vs. from seed/pollination) requires keen dexterity, a sharp eye and an even sharper knife.

"It is surprising," Luther Burbank wrote in 1916, "how little is known by the general public about the propagation of fruit trees."

Contrary to common belief, a Golden Delicious tree does not come from a Golden Delicious seed. Each seed from a single apple will provide different apple varieties, varying in size, color, shape and flavor. The only way to get an exact variety of apple is to either graft or bud a part of the parent tree to rootstock and grow a "new" tree — essentially a clone of the parent tree. Fruit trees grown by Stark Bro's are budded or grafted so the young tree will produce "true-to-name" fruit.

The grafting house in the 1930s. Almost all trees are budded now.

The art of grafting. A skilled grafter can graft 2,000-2,500 trees per day.

Grafting is the precise process of attaching the scion of one tree to the rootstock of another. The above-ground part (the scion) will produce the branches, leaves, flowers and fruit. The lower portion (the rootstock) produces the root system.

The grafter makes sloping cuts on both the scion and rootstock then cuts a notch into the face of each sloping cut so that the two pieces fit snugly together. The graft union is then wrapped with adhesive tape or rubber strips to hold it together, and is labeled for identification.

Budding is very similar to grafting, but instead of working with a twig that has multiple buds, only one bud is used. Since about 1991, most propagation is done by budding because it's much faster and easier. Stark Bro's grafts in the winter into spring, and buds in the summer, when growth has slowed down.

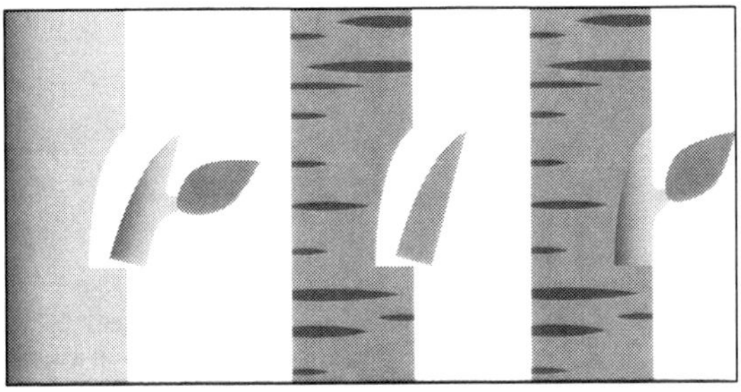

Budding only requires attaching one bud from a parent tree to a compatible rootscock.
Graphic: Will Carson

One might guess that a good budder could do up to a hundred trees per shift. Amazingly, Kenny Harrison budded 3,000 trees one day. But it's his wife, Sue, who holds the incomprehensible record: 4,750 trees in *one day*.

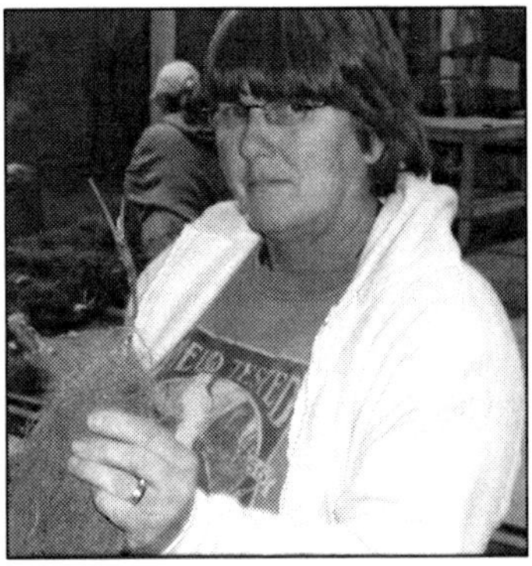

Record-holder Sue Harrison

Budding is still tricky, but is much simpler than grafting. Workers cut shoot buds from parent trees of the scion variety, and bring them back to the nursery row where they are slipped into the bark of the rootstocks that were planted earlier that year.

These buds are wrapped with rubber or plastic bands until they callous and fuse with the bark of the rootstock. The top of the rootstock, above the scion bud, is cut back early the next spring, and the scion bud grows to form the top of a finished five- or six-foot tree in a single growing season. If a variety becomes a success, a number of trees are set out in a "scion orchard" as a source of propagating-wood for next year.

Luther Burbank's amazing 200-graft cherry tree

Other plants, like ornamentals and berries, are propagated by cuttings, which are rooted in sterilized soil under plastic covers. The foliage is automatically misted at timed intervals until roots develop. The covers are removed and the cuttings grow into full-fledged plants ready for transplanting.

10

The First Catalogs

About Clarence, and the Family Feud

The Fourth Generation Comes Up

The Wily William P.

10 | CLARENCE AND THE BROTHERS

1894-1913

The brothers' exciting new association with the world's greatest horticulturist fueled their collective imagination. They could not wait to start selling Luther's botanical marvels alongside their extraordinary Delicious apple. The prospects were tantalizing.

The First Catalogs

The company's first catalog, produced by Clarence Stark in 1894, contained 16 pages of type sprinkled with pictures of fruits, orchards and orchard owners. The cover consisted of three photographs: 300 workers at the nursery, lined up for payday (right); another of men at work in the grafting house and a photo of a 25-acre rose field. This marked the first time a photograph was used in a Stark catalog.

Stark Bro's first-ever catalog is long gone, but this is one of the photos that was on the cover.

By this time, the business had branched out to include peaches, apricots, plums, pears, cherries, roses, and ornamental trees and plants, but the principal product was still apple trees. The Stark Bro's catalog listed each variety with a short, factual description; first-choice apple names were in bold capital letters, the second-choice sorts were in lower-case type. Clarence's copy was at the very least, honest.

"Second-choice sorts are rejected because [they are] either worthless or superseded by better sorts of the same season. We grow some second-choice sorts because many planters know and want them and won't be convinced, for a time at least, that the others are better. When any sort proves itself in every way superior to another, away goes the inferior sort into second-choice lists, or if too seriously lacking, is wholly discarded."

Regarding a first-choice apple, he waxed poetic:

> *"We have never eaten an apple, which, in perfection of quality, is so nearly ideal," he wrote of the Akin apple. "Besides possessing excellence undreamed of in the philosophy of the apple markets, it has just the right size and brilliant color not only for hotel and fancy trade, but for export."*

Clarence McDowell Stark was an eloquent, and persuasive, salesman.

The early catalogs were homespun and conversational, and clearly written with the presumption that on the farm, winter nights were long and reading material was scarce. It has been said that for several generations, the Bible, the Sears Roebuck catalog and Stark Bro's catalogs were the only pieces of permanent literature found in many farmhouses.

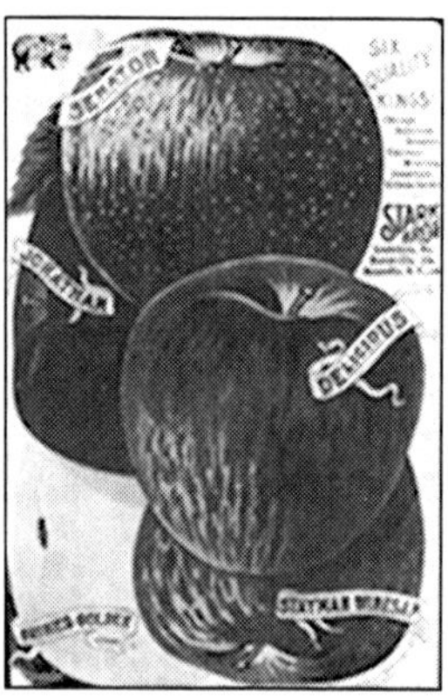

Stark Bro's first color catalog, 1896

The 1896 edition of the catalog introduced color illustrations of Stark fruit, which were "painted from nature." It also contained a portrait of William Shakespeare (whom Clarence Stark liked to quote) to illustrate a short piece on the nurseryman's art:

"This is an art which does mend nature," said William Shakespeare in 'A Winter's Tale,' writing of the nurseryman's budding and grafting skill. "Truly herein lies the real art of the nurseryman, not in producing strawberries and so forth. These require no art nor skill to raise, but grow just as the grass grows in the field."

The detailed color catalog, in tandem with the transcontinental railroad, put the Stark name on the map. The Stark nurseries now offered nearly 50 varieties of apples, and they were shouting it from the rooftops. An ad appeared in the first issue of *Collier's Weekly Magazine,* making Stark Bro's one of the first advertisers to appear in weekly publications read throughout America and abroad.

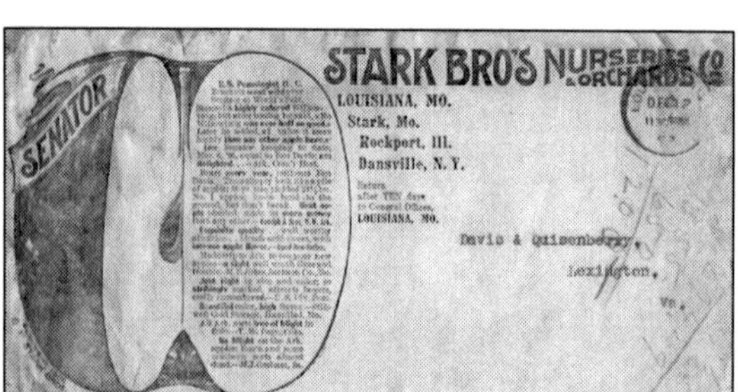

An 1896 advertisement/postcard

As the business continued to expand, the Starks felt a need for another member of the team, a sort of "quarterback" to synchronize the operation. They brought in their cousin, Eugene Stark. He was the son of their uncle, Thornton Grimsley Stark, who, for a time, had been active in the nursery with their father, William. They gave Eugene a 10 percent interest in the company in return for his farm near Louisiana.

Eugene functioned as secretary and office manager, but he was even more important as a goodwill ambassador. Eugene knew how to handle people, and he also helped to coordinate the work of the other brothers. Eugene had also served as county Judge of Pike County, and later his political experience and his wide popularity won him election to the Missouri Senate.His workload at the nursery office combined with his position as State Senator was visibly sapping him, and as his popularity grew, friends urged him to run for governor. The stress and schedule were too heavy a burden, and after a major operation, Eugene died in 1909 at the young age of 44.

Clarence's very expensive color catalog for the 1904 World's Fair

By 1902, the company's capitalization had increased to $1million, with 5,000 shares of common and 5,000 shares of preferred stock. Not more than a dozen shares were outside of the immediate family, and those were owned by retired employees of long standing, or by distant relatives. The Stark men kept their financial affairs close to the vest, as well as their family affairs. Clarence, now 47 and brilliant as he was, seemed to slip into odd behaviors at times, doing and saying things no one could quite understand. Everyone knew geniuses were a little "different;" he was just being Clarence, they said.

The World's Fair in St. Louis was officially opened in 1904 by President Theodore Roosevelt, who visited Louisiana, Missouri as part of the trip. This national event gave huge exposure to Stark's creations. For this special occasion, Clarence had more than 100,000 eye-popping color catalogs produced, and plastered the fairgrounds with them. This was the first time people had seen a sensational horticultural catalog of this size and scope, in color. The shrewd campaign paid off mightily, bringing in an avalanche of new customers.

President Theodore Roosevelt visits Louisiana, Missouri as part of his trip to the 1904 World's Fair. In the inset, Edgar Stark is speaking with the President.

About Clarence, and the Family Feud

The grand success of the World's Fair was now two years behind them, and with each passing month, it became painfully obvious to Edgar and William P. that Clarence's "odd behavior" was much more than a quirk. His mental health was deteriorating quickly.

Newspaper reports place him in a San Antonio, Texas sanitarium in 1906, but it is possible he was there before that. He wandered away from the facility on April 1st, 1906 and was not found until the next Wednesday, when a local sheriff discovered him. He was returned to the facility and remained there for several more years.

The story of his 1887 trip to Colorado for a "tuberculosis cure" might possibly have been in response to the early onset of Alzheimer's or dementia. There are no records to indicate that he ever suffered from TB. It is speculation, but one must remember in those days that mental illness had a strong stigma attached, and a prominent family like the Starks would likely want to distance themselves from it, even if that distance were as far away as Texas.

By all accounts, Edgar was the one who picked up the ball and managed Clarence's presidential duties; he was next oldest in line. William P., taking umbrage beyond reason, chose to sulk.

Clarence, out in the orchards. Date unknown.

By 1909, it became painfully clear that Clarence would need to be formally replaced as the head of the company. William P. was becoming increasingly agitated with the fact that Edgar was sitting in the president's office instead of himself. In a note written to Edgar in the spring of 1909, he said he "was through" and offered to take a dollar amount for his share, to which Edgar agreed. William then "crawfished out of the deal," according to Edgar's personal papers. And so the stew simmered.

At the annual meeting that June, William P. demanded that his son, William H., be made a Director, Assistant Treasurer and VP, and that his son-in-law be made Secretary (he had been with Stark Bro's less than a year). This would have given William P. half of the Board's voting power. Edgar was incensed.

In the new year, William P. proposed to Edgar that the two of them buy

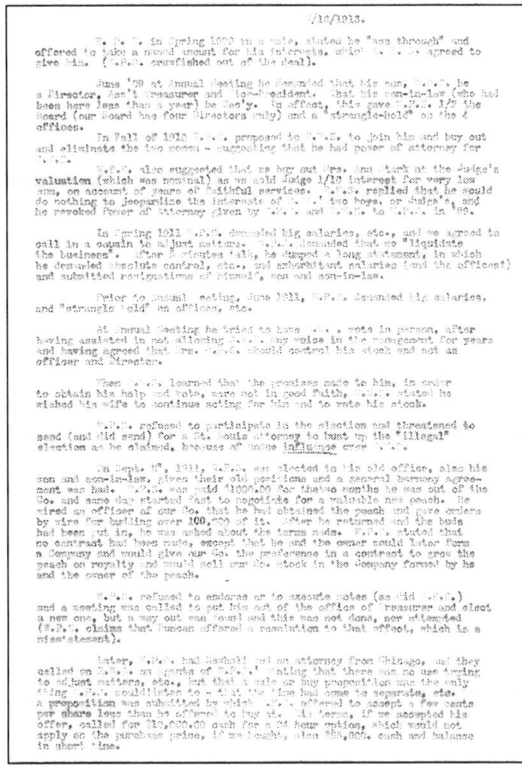

A portion of Edgar's personal notes, written in hindsight about the William P. "situation"

out several of the other major shareholders, including Clarence's wife. Even more infuriated, Edgar revoked the power of attorney given to William P. in 1889, vowing never to jeopardize the interests of Clarence's two boys or that of Judge Eugene. The intensity of the feud was becoming volcanic.

The following spring, William P. strode into Edgar's office with his outrageous demands: he wanted an exorbitant salary increase, and ordered Edgar to "liquidate the business." He then dumped a long statement on the desk, demanding absolute control, a ridiculous salary for his son, and the office buildings, along with written resignations for himself, his son and his son-in law.

Edgar was floored by the audacity of his brother. While he most certainly would have celebrated the end of William P.'s incessant whining, Edgar did not accept the resignations. He was compelled by what he considered to be the "sanctity of family tradition" to keep the business alive.

On the personal front, there was troubling news. Clarence, after spending years in a mental institution, had been declared legally insane by a Texas probate court. His wife, Lilly Stark, had him moved to the Maplewood Sanitarium in Jacksonville, Illinois (with the court's permission). At least he had been spared the ugliness of his brothers' years of battling.

It was in May of 1911 that Edgar and the cantankerous William P. did agree on one thing: that Clarence be "eliminated from any actual control of any branch of the business whatever," according to Edgar's personal papers.

The truce lasted about as long as a sneeze. At the annual meeting in June, William was back to his old list of demands. He refused to participate in the election of the officers, and threatened to send a St. Louis attorney to "bust up the illegal election," as he called it.

For the sake of the business — but not until September 2 — William P. was elected to his old office of Treasurer, and his son and son-in-law were given back their old positions. A "general harmony agreement" was had. Edgar wrote a check for $1,000.00 for the two months William was out of the company.

That same day, William headed east to negotiate for a valuable new peach. He gave orders by wire for budding more than 100,000 of them. After he returned and the peach buds had been budded, Edgar asked his brother about the terms of the deal. William P., with gall matching the size of his ego, explained that there was no contract. He and the owner would later form a company and give Stark Bro's preference to grow the peach on royalty. To cap it off, William P. offered Edgar stock in his newly-founded company. It was blasphemy.

Young Edgar W. Stark

Rumor has it that Edgar threw a paperweight across the room and shattered a glass pitcher. It was a falling-out of gigantic proportions.

Grasping fairly quickly that he wasn't going to come out on top, William retaliated by refusing to do his Treasurer's duty to endorse notes (as did his son), bringing the company to a standstill. An emergency meeting was called to put William P. out of office and elect a new Treasurer. Somehow, this was not done, and so, quite unbelievably, they entered into General Harmony Agreement #2.

Later, Edgar heard from two of William's attorneys, insisting that a sale or buy proposition was the only thing in play, and that the time had come to separate. Said Edgar, "the harmony agreements went where the woodbine twineth," meaning into the ether. Edgar might have been irate, but at least his sense of humor was intact.

Leadership had all but disappeared what with all of the lawyers and fighting and intra-family tension. Edgar decided to call in the cavalry, and deftly managed to get his nephew, Lloyd Crow Stark, sent home from the Navy to fill a "critical management vacancy." Old family friend, House Speaker Champ Clark, interceded on Lloyd's behalf, and he was allowed to resign from the service. Lloyd was instantly made VP and General Manager.

Paul Stark Sr., Lloyd's younger brother, got on board the following year. One of the first in the Stark line to attend a formal school of horticulture, he graduated from Cornell University in 1912. Paul went on to do important post-graduate work in fruit-growing and nursery investigation.

The two brothers, along with their uncle Edgar (who had supervised production and finances for many years), began to accept more management responsibility.

The Fourth Generation Comes Up

One of the first major decisions made by brothers Lloyd and Paul Sr. was about advertising. World War I had just started overseas, and worries started bubbling up about the American economy, even though the U. S. did not enter the war until nearly three years later. Some in the company wanted to retrench and cut down on advertising, arguing that the company was already well known.

A Stark barn ad in Pennsylvania

"But we remembered something our father had always preached to us," said Lloyd. 'Boys, every dollar spent for good, sound advertising and promotion will eventually bring two dollars rolling back to you. Never stop advertising — do this in good times and bad times, and the business will grow and prosper.'"

The Stark Bear Barn, in Louisiana, MO

The brothers not only kept advertising, but increased the budget. Newspapers and magazines continued to display Stark Bro's ads, as did the sides of barns and buildings across the country. It was hard to avoid that big black bear logo informing everyone who drove by that "Stark Trees Bear Fruit."

The Wily William P.

In a not-so-startling turn of events, on January 1, 1912, William P. and his son opened William P. Stark Nurseries in Chester, Missouri, which was housed in the building of the old Mutual Wagon Factory. He pushed very hard to have the town renamed Stark City (after himself, of course), and was actually successful.

He adopted a trademark that prominently displayed the name "Stark Trees." Unfortunately, he must have forgotten that he was still an officer of another company until May 1st (because his resignation was never accepted). He unscrupulously drew a salary from Stark Bro's all the while.

To add insult, William P. and his son each reneged on financial obligations owed the company, to the tune of about $5,000. From Edgar's personal papers, it sounds as though he could barely be in the same room with the two of them.

William P. was finally, and unceremoniously, un-elected Treasurer on July 1, 1912. According to Edgar, all ties with him were "severed," a genteel way of saying he was booted out — of the company, and, for all intents and purposes, the family.

William P. Stark's very similar logo and audacious letterhead (below)

Said Clay Stark Logan, Edgar's great-grandson: "This is a good case study of governance in family businesses."

And so the exiled brother swaggered off to his own nursery, secure in the knowledge that he would outdo Edgar. He was paradoxically philanthropic around town, and to this day, there is an imposing stained-glass window at the Louisiana First Presbyterian Church with his name prominently etched into a section of red glass. William P. was, if nothing else, a very competent self-promoter.

After a while, William P. put forth the absurd claim that Edgar was a "shadow" and "pretense" in Clarence's absence. He could not let it go. William P. sued his own family's company in March, asking for the appointment of a receiver. The lawsuit accused Edgar of fraud, incompetency and mismanagement (among other things). A change of venue was asked for, and granted (Starks were everywhere in Pike County, and could hardly be counted on to betray their kin), and so the case was moved to Audrain County, Missouri.

The case went to trial that fall and the Honorable J. D. Barnett denied the receivership. The judge found Edgar to be "a very efficient element in the upbuilding of the corporation." In fact, Edgar was a very astute businessperson, as evidenced by the multiple trusts he set up — some of which lasted until the early 1980s.

And so William P. slithered away ... for a while.

A Tale of Two Stark Cities

Stark City, Missouri, once known as Starkdale, came into being when Stark Bro's Nurseries & Orchards Co. leased some land there from Fieldon Estes for tree propagation. At its bustling height, around 1900, Starkdale boasted a general store, a school, a railroad station, a post office and about 100 families. The post office quit operating in 1904 but remains as a historical marker, and that is about all that's left of Starkdale. The population was 139 as of the 2010 census. It was formerly called Chester, but was renamed in the 1930s after a lot of pushing by William P. Stark, who had started his own nursery business there and wanted to capitalize on the golden family name. William P. Stark Nurseries was a going concern in Stark City as well as in Newtonia, its next-door neighbor.

The old Stark trolley truck, circa 1924, used to drive around the grounds, giving tours to the public

A horticultural Who's Who visited Stark Bro's to see the Starking® Delicious

The Stark women employees pose on the front lawn of Gov. Lloyd Stark's mansion. He used to be their boss.

The 1928 International New Fruit Fair, which unfortunately did not reveal another Delicious apple

11

Mr. Mullins Sends a Sample

At First, A Golden Flop

Delicious Goes Worldwide

11 | THE TRAIL OF THE GOLDEN APPLE

1914-1919

The Starks learned a very valuable lesson from their discovery of the Delicious apple, and that is: while you can perform wonders with cross-breeding, nature can come up with some pretty stunning surprises of its own.

Growers regularly encounter "chance mutations," most of which amount to nothing — but every so often, a good variety pops up and gets sold for an impressive sum. Stark Bro's still accepts samples from growers hoping to "hit the jackpot" because one never knows what might be in the box. That policy has rewarded the business more than once.

Mr. Mullins Sends a Sample

One day in April, a young Lloyd Crow Stark opened a small box of sample apples that had come in by parcel post, addressed to Paul Sr. It contained three yellow apples; nothing earth-shattering, as he had seen 'Grimes Golden' and a few other yellow varieties before. None of them had ever garnered much attention because they were difficult to keep, and over time, became flat-tasting and mealy.

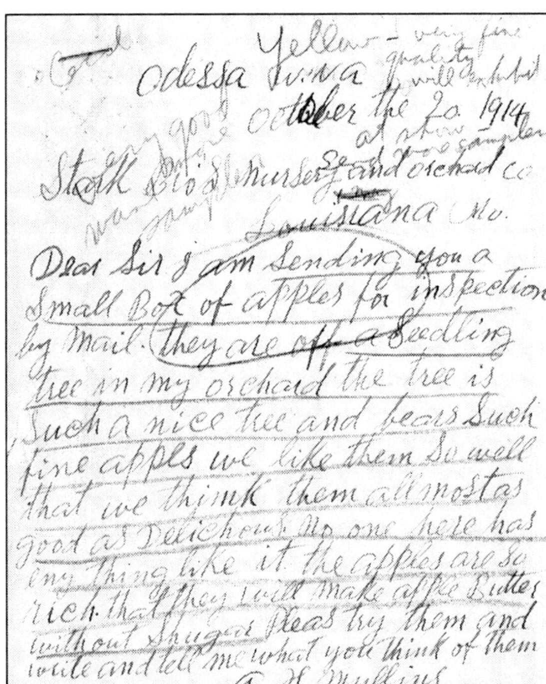

Odessa W.Va October the 20. 1914
Stark Bro's Nursery and orchard Co
Louisiana Mo.
Dear Sir I am Sending you a Small Box of apples for inspection by Mail. they are off a Seedling tree in my orchard the tree is Such a nice tree and bears Such fine apples we like them So well that we think them allmost as good as Delicious No one here has any thing like it the apples are So rich that they will make apple Butter without Shugar Pleas try them and write and tell me what you think of them
A H Mullins

The original letter requesting Stark Bro's to sample what came to be known as the Golden Delicious apple.

Paul happened to be wandering by, and stopped in out of sheer curiosity. He had been keeping an eye out for another apple to rival the popularity of his father's discovery, the (red) Delicious, but was sure this yellow apple wasn't it. Lloyd invited him to share a taste anyway; he cut two slices and they each took a bite. Contrary to some stories, Paul did not say "Eureka! This is it!" He and Lloyd simply looked at one another in silent astonishment.

This was no Grimes Golden — this was spicy, crisp and juicy, an impossible state of being for an overwintered, parcel-posted Grimes apple.

The box had come from a farmer named Anderson H. Mullins of Odessa, West Virginia, a spot so obscure it didn't even appear on a map. In 1905, Mullins had noticed an odd seedling in his orchard, one sporting large yellow apples, and was surprised that it was still bearing after nine years. He was so proud of his "Mullins Yellow" that he sent three samples to the only other man he knew could appreciate them: Paul Stark Sr., the leading apple authority of his time.

Famous horticulturists gathered to study the Golden Delicious apple.

Farmer Mullins received a polite and complimentary letter from Mr. Stark, remarking on the apple's pleasant spice and suggesting that a visit might be in order to verify the uniqueness of this new variety.

And so that autumn of 1914, Paul Sr. set off for West Virginia, nearly 1,000 miles from Louisiana, on what he called "The Trail of the Golden Apple." He could reach Odessa only via narrow-gauge railroad, after which he'd have to complete the final 20 miles on horseback.

Stark gamely rode through the mountains and finally came upon a mailbox with "Mullins" scrawled on it. He knocked on the cabin door repeatedly, but no one came. About to give up, he decided to look around back, where he found a small, disregarded orchard dotted with some gnarled apple trees. A chance glance upward at a craggy hill met his eye with a glint of gold. It was the parent tree, dripping with big, yellow fruit.

Paul described the scene: "There was one tree with rich green foliage, looking as though it had been transplanted from the Garden of Eden, its boughs bending beneath a tremendous crop of great, glowing apples." He was heartily clamping down on one of them when a ragged man seemingly appeared from nowhere.

Without introducing himself, Stark reportedly said, "That's some apple."

"Name's Mullins," the man replied, apparently possessing better manners than the wealthy apple researcher. "That tree just growed there," Mullins explained. "I sent you some apples from it."

The two went inside and hammered out an agreement. Stark paid Mullins $5,000 for the rights and a 900 square-foot piece of land around it — but no royalties — and the gorgeous golden apples became the property of Stark Bro's Nurseries & Orchards Co.

The original caged Golden Delicious tree with Paul Stark Sr. standing at the base of the hill.

A woven-steel cage was built around the tree so no one could steal any cuttings; the cage was also equipped with a burglar alarm. It was a sensational tourist attraction.

For 30 years, the company sent Anderson's nephew, Bewel Mullins, $100 a year to look after the tree and keep records on it. Bewel also sent scions back for propagating, which would create thousands of clones of the original tree.

The Mullins tree's last bearing season was 1958. Today, the only evidence of that golden goose of a tree is a historical marker out on Route 1 near Porter's Creek. Though the tree stands no more, one has only to visit the supermarket to see the lasting impact of the Golden Delicious, now the state fruit of West Virginia and one of the best-selling apples of all time.

Historical marker at the site of the Golden Delicious discovery

The discovery, introduction and assertive advertising of this apple — and its erstwhile "cousin," the by-now trademarked 'Red Delicious' — crowned Stark Bro's as the uncontested king of the fruit-tree industry and proved once and for all that money, does indeed, grow on trees.

Paul Sr. made the long trip back to Missouri with an armful of scions, and it wasn't long before the nursery was commercially producing Stark Golden Delicious apples, the first yellow apple with taste, size and keeping ability. It was wildly popular and would finance the nursery for decades to come.

Later research at the Stark laboratories determined that the tree was a nature-bred chance seedling, a product of a 'Golden Reinette' seed, a French favorite, pollinated with either Grimes Golden, or possibly 'Golden Russet.' Some pomologists hypothesize that it could have even been Johnny Appleseed's doing, accidental seedling that it was.

At First, a Golden Flop

Interestingly, the marketplace had a tepid reaction to the Golden Delicious — much like it did to the odd-shaped Red Delicious — at first.

"No one's going to want a yellow apple," said the buyers. Again, Stark included bundles of free trees with commercial orders, expecting money to pour in a couple of years down the road.

But this time, the response was almost non-existent. Mystified, men from Stark went out to the commercial orchards, armed with samples of this great fruit, to investigate. They discovered that because those "yellow apples" were considered worthless, they had been stuck in poor soil and neglected, deemed unworthy of attention. No wonder they fared poorly.

Golden Delicious, a heavy and early bearer, just 2-1/2 years after planting, broke records in 33 states.

Once the growers tasted the result of a cared-for tree, all doubt was erased and the Golden Delicious orders were soon backlogged. Between the unequaled flavor of both the Red and the Golden, there was such intense public demand that "Stark Delicious" became a standing order for buyers.

Delicious Goes Worldwide

"American missionaries were among our first agents in foreign lands," said the late John Stark Logan, president from 1964 to 1984. "If they found no

The Queen's Favorite Apples

If you should ever visit the Louisiana Area Historical Museum, you can see a lovely old, larger-than-life-size portrait of Queen Marie of Rumania holding a Stark Golden Delicious apple. The painting was made from a photograph, which the Queen at first refused to allow. But when she saw how much members of her royal party enjoyed the new variety of apples given to her by Stark Bro's, she summoned the photographer back. The Queen's picture and the story of how she was won over were included for many years in the Stark catalogs.

In addition to this painting, the museum houses "The Stark Room," a wonderful collection of Stark Bro's memorabilia, old tools, papers, photos and even some of Luther Burbank's notes and letters.

apples where they were stationed, they would send for our trees. Many of them persuaded public officials that apples would improve the diet of the people, and so agricultural representatives began coming to our nurseries to learn how to develop commercial orchards."

Stark trees have been shipped to every country in Latin America. In Europe, France and Italy plant upwards of 50 percent in Delicious types, which are also grown in Belgium, Holland, West Germany, the Baltic countries and even some areas of southern Russia. The principal apple in the Middle East is the Delicious. Lebanon even issued two postage stamps to honor the famous apple. Add to the list: India, Japan, Australia, New Zealand, Tasmania and South Africa, and it becomes increasingly difficult to find a country where Delicious is not grown.

Golden Delicious won a prestigious Wilder Medal, the "Nobel Prize of the Fruit World"

In 1964, Paul Stark Jr. estimated that 24 million trees had been grafted from the original tree. In 2015, clones grow on five continents and produce more than 200 billion pounds of apples each year in the U. S. alone.

Golden Delicious won the "Nobel Prize of the Fruit World" in 1919, the coveted Wilder Medal. Golden Delicious is as close as it gets to immortal in the apple industry.

12

Dear Clarence

An Infringement To End Them All

The Reign of Bud-Sports Begins

The Plant Patent Act

12 | THE WORLD WAR, AND THE OTHER WAR

1914-1932

While Golden Delicious was booming, World War I was literally exploding. When the official war was declared in July of 1914, Lloyd C. Stark, then President of the American Association of Nurserymen, resigned that position and applied for reinstatement as an officer in the Navy. He was offered an Army commission, and went to France with the 80th Division as a Major in the 315th Field Artillery.

Edgar comfortably assumed Lloyd's place, and with a significant amount of help from his nephew Paul Sr., they managed the nursery operations.

Business was lean because of wartime restrictions and shortages; Washington had decided that nursery stock was non-essential, and nurseries were not allowed to ship by rail, a precious mode of transportation at the time. "It threatened to break us," said Paul Sr. Fortunately, the rule was later relaxed after the authorities realized the food-supply and health advantages of fruit-bearing trees and plants. Lloyd returned from active service and resumed his duties, and Paul Sr. moved into the Treasurer position.

Dear Clarence

All this while, lovable, quirky Clarence, after being legally declared insane in 1911, was still in the Illinois "sanitarium," as mental institutions were called then. Tragically, on May 30, 1914, Clarence "came to his death from a self-administered overdose of poison, taken to allay nervousness and produce sleep. Mr. Stark was found dead in bed," according to the coroner's inquest report.

The one brother who understood the power of advertising like no other, was gone. Sadly, Clarence died before he could witness the enormous success of Golden Delicious, but his life had been plenty blessed. Edgar wrote of him:

> *"The fruit tree knowledge of four generations of Starks was concentrated in this talented brother of mine. (He) devoted his life to introducing the best varieties of fruits he could find. I often heard him say: 'The nursery business has always meant more to me than a mere means of livelihood. Whenever I can persuade a man to plant a tree that will bear better fruit, I feel I am doing him a favor for which he will thank me when harvest time comes'."*

Edgar formally succeeded Clarence as president, eventually spending 60 years in the business. He was a very able businessman, and in modern times he would have been labeled as an "efficiency expert." Edgar was one of the first in the industry to introduce a "production line" in the business, and he was known to have an eye for recruiting and developing new talent.

A case in point was a young man named Homer L. Reed, who displayed the qualities Edgar Stark valued so deeply. When the production manager at the nursery died, Edgar selected Reed, then only 26, to fill that spot. Homer would serve in that capacity for more than 50 years. Edgar and Homer, as a team, made game-changing improvements in nursery operations: they built huge packing houses and brought the labor — which had always been outside, in all kinds of weather — inside. More importantly, they developed the mechanical tree-digger, a milestone in nursery production. Homer's visionary inventions, and Edgar's ability to make them a reality, forever changed the industry.

Homer's mechanical tree digger

A later version of the two-row cultivator Ernie Elder developed. Stark Bro's still uses this machine.

After a half century as production manager, Homer Reed retired, but he was kept on full salary as a "consultant" and devoted himself to the invention of time- and labor-saving devices, now commonplace in commercial nurseries everywhere. Homer was replaced by Ernie Elder, who was every bit the mechanical whiz Homer was. It was said by many that "he could just look at a job, and then build a machine to do it."

An Infringement to End Them All

As if World War I and the passing of Clarence weren't upsetting enough, William P. was elbowing Edgar once again. He was flaunting the Stark name with impunity, hawking his nursery stock like a carnival barker, offering "genuine Delicious trees ... at grower's prices," trying to undercut his brother.

Edgar decided that William P. would shortly become the defendant in a long, ferocious lawsuit alleging trademark infringement (displaying the words "Stark

Trees" in his advertisements) and unfair competition (using the name "Stark"). Edgar also wanted damages. The proceedings would clearly, and permanently, color William as the black sheep of the family.

Said Edgar in notes written about 12 years after the lawsuit: "[William's] head was turned with the Company's success. He minimized everything the others did and magnified everything he did. It was a case of too much flattery — ending in a swelled head." The comments were really quite charitable, considering.

The landmark case was filed in District Court and went through multiple appeals, ultimately ending up in the Supreme Court of the United States. The suit would spend six long years traveling through the judicial system.

Despite the legal spitting contest, life and work went on as usual. Delicious apples were still big sellers, but other important varieties such as 'McIntosh', Jonathan, Rome Beauty, along with 'Elberta' and 'J. H. Hale' peaches, were gaining prominence.

On the home front, Paul Stark Sr. eloped a few months before his planned wedding to Theodosia Barnett. The local newspaper alerted the town that the marriage occurred "without informing the family."

Meanwhile, the contentious lawsuit was winding its way through the system when rascally William P. ran a full-page ad in *The Country Gentleman* to tout "his" remarkable new J. H. Hale peach. The condescension in his copywriting managed to insult potential customers:

> *"Well-informed quick people have planted it by the thousand; well-informed slow people are experimenting with it; the misinformed are still planting the less dependable, less profitable varieties."*

He even managed to disparage his brother's product in the same ad:

> *"The best varieties are the greatest money-makers. Witness the Elberta peach. Now there is a greater [peach] than Elberta. The J. H. Hale has the same good qualities, but to a greater degree."*

Under William's picture, the caption (also written by him) reads:

> "William P. Stark. Recognized as a leader for his improvements in nursery methods and in securing better horticultural laws. He had the foresight and courage to propagate and advertise the J. H. Hale peach and the Delicious apple."

None of which is true, but that never stopped William Pharr Stark.

Full-page ad for Wm. P. Stark Nurseries in The Country Gentleman, *1/27/1917. Below are Clarence's "genuine" seals.*

Stark Bro's launched a counterattack by running a full-page ad in the same magazine, boasting, "J. H. Hale Peaches from Stark Bro's Yielding Record Crops Everywhere." (Note the underline of "Stark Bro's," in an effort to distinguish itself from Wm. P. Stark Nurseries.)

The copy goes on to explain that Stark Bro's had been growing "genuine" J. H. Hale peaches since 1911 (insinuating that William P. was fraudulently claiming the variety for himself). This would have been a dandy Twitter war, had these events transpired today.

In 1917, Josiah J. Hazen assumed control of W. P.'s company, for reasons undetermined. In 1925, A. E. Weston went to work for Hazen as the general manager. (A. E.'s grandson, Joe Weston, Jr., would later end up as the head of Stark's Fruit Tree Center division, in true small-town fashion.)

William P. Stark Nurseries was reorganized and the name was changed to Neosho Nurseries Co. The restructuring did not include William P., who was ousted by Josiah Hazen and other investors, although he retained a minority share. Lucy, William's wife, "pestered me annually for the dividends," said Joe Weston, Jr. grandson of the owner.

William P. abruptly left the state, taking his colorfully-named secretary, Peachy Ransom, with him. He became the general manager of the Cottage Gardens Nurseries in Eureka, CA. Peachy worked for William P. well into his seventies.

As for the infringement lawsuit: after six grueling years of litigation, the Supreme Court of the United States found in favor of Stark Bro's on both the trademark infringement and unfair competition allegations. Damages were limited to the date of the lawsuit filing. William P. Stark was enjoined (in absentia) from those practices and the use of the post office address, "Stark City."

The other Stark brothers, after a conference with Josiah Hazen, agreed to accept $7,500 in place of the damages and profits awarded them by the Court. This amount was paid by Neosho Nurseries and injunctive relief was awarded to Stark Bro's against further infringements and unfair trade, in perpetuity.

William never uttered a peep about the outcome of the suit to any of the brothers. There is some evidence of him living out his life in Philadelphia, but no one knows for sure what happened to the wily William P. He died in 1943.

The Reign of Bud-Sports Begins

After years of national promotion, the Delicious apple became a leading commercial variety, and by 1922, the crop was valued at about $22 million. Golden Delicious was well on its way to achieving the same stature. It was about time for something new.

A bud sport (chance genetic mutation of a tree) can affect the color, size, shape, quality, time of ripening or even the size of the tree itself. All rootstocks onto which scions from a "freak of nature" limb are grafted will produce fruit of the new variety. Many believe bud sports occur because of natural radiation from the sun, but that is still up for debate.

The Starks realized the possibilities of bud sports when they first discovered the limb which gave the world the Starking® Delicious apple, and later with the discovery of Starkrimson® Delicious and the Starkspur® Golden Delicious. These semi-dwarf spur-type trees bear heavily all through the tree, not just on the limb-ends. They double the production per acre, which means a much lower cost-per-bushel, and higher profits. Spur-type trees clearly solve many problems for orchardists.

In June of 1923, a New Jersey orchardist wrote to the Starks to inform them that one limb of his Delicious tree was "acting strangely," producing red apples, while the rest remained green. A chance mutation that made the apples color up earlier had also given them a deeper, redder color, and the stripes were gone.

Starking® Delicious turns brilliant red three weeks earlier than Delicious.

Customers buy with their eyes, and the Starks knew it; this was worth investigating. Paul Sr. hurriedly went to New Jersey and paid Lewis Mood a whopping $6,000 for that bud-sport limb. It was trademarked Starking® Delicious. From this one limb, many millions of trees in orchards all over the world trace their origin.

Lewis Mood's original Starking® Delicious bud-sport limb

Word got around, and soon *The Gettysburg Times* reported that more than 500 horticulturists had gathered at the orchard to discuss the "freak bud" that produced "the marvel apple of the age."

Starking® Delicious was unanimously awarded a Wilder Medal, and now the Starks had two. The variety quickly became the best-selling apple in the world, and has enormous significance in the fight against hunger in poor countries because of its compact size and heavy-bearing traits.

The first conference ever held to evaluate the possibilities and impact of bud sports on the fruit industry occurred in Monroeville, New Jersey, in 1925. The nation's leading fruit scientists from Washington, D. C., a large group of university horticulturists and industry leaders attended to inspect this new bud sport of Delicious. The conference and the interest it aroused have been a major factor in fruit improvement in the past 90 years.

Starking® Delicious (right) colors earlier than the original Delicious, but does not ripen earlier.

Bud sports are now recognized as a very important means of improving trees and plants of well-known varieties with which the public is already familiar.

Is Redder Better?

Once in a while a branch will bear redder, or longer, or pointier fruit (all more eye-appealing and therefore more sellable); it was logical that whole trees of such fruit would sell better, so growers would propagate that branch to make whole trees with redder, longer, pointier fruit. The grocery stores care about looks and shelf life, but the backyard gardener and truck farmers grow for flavor.

The further you get away from the original product, the better — or worse — it can be. The Red Delicious has been mutating and re-propagated so many times for appearance, and not taste, that no one walking around today will ever know how delicious that original Red Delicious really was. For sake of "pretty," growers have propagated the flavor right out of the apple.

To make matters worse, the fruit of some of those strains colors up as much as two weeks before harvest, which could result in some pretty terrible-tasting apples if a dishonest fruit grower were trying to take advantage of that appealing red color. Sad to say, the fruit that so excited Clarence Stark's palate in 1893 is not at all what is sold in today's produce departments.

The Plant Patent Act

Not long after the discovery of Lewis Mood's Starking® Delicious, Luther Burbank passed away. It was a personal blow to the Stark brothers, for they lost a friend and magnanimous collaborator.

A high bar was set; they were now in possession of Luther's seeds, plants and notes, as he had wished. These small items represented big shoes to fill, but they also served as deeply embedded inspiration to do great things in the void of "The Wizard of Horticulture." It was only fitting that the Starks begin work to protect the discoveries of fruit developers like Burbank, and themselves.

Some say the Plant Patent Act (PPA) was far too long in coming. If it had been in place earlier, then Stark Bro's would not have had to erect a cage around the Golden Delicious tree. That cage only stopped people from propagating it; however, once Stark Bro's started selling trees, those trees could be used by anyone to propagate new ones — because they did not hold a patent. A disparaging problem, indeed.

An original breeder's one shot at financial remuneration was to sell it out of the gate in limited quantities, and at a very high price. Soon thereafter, presuming the variety was a good one, the market would be flooded with the variety and the price diluted to virtually nothing, another defeat for innovation.

It was Paul Stark Sr. who was the driving force behind the PPA. Another horticulturist-nurseryman who was essential was Archibald Augustine of Bloomington, Illinois. He had served on the National Committee for Plant Patents for many years, but little had been accomplished because, frankly, Congress didn't understand the issue very well.

When Augustine was elected president of the American Association of Nurserymen in 1929, he went to Paul Stark (then chairman of the National Committee for Plant Patents) and asked him to lead the fight in Washington. Clarence Stark had clamored for this decades ago, and now Paul Sr. was getting close to making it a reality.

Plant breeders, Stark's committee argued before Congress, created new varieties of plants at great effort and expense, but without legal protection of any kind. The agriculture industry asked only for the same consideration that was granted to other creative endeavors: mechanical inventors, authors, composers and such.

Some of these "protected" people even volunteered their support to the Committee, including Thomas Edison, a friend of the late Luther Burbank. Edison's telegram to Congress strongly urged adoption of the legislation. He also testified before Congress, stating, "This [bill] will, I feel sure, give us many Burbanks."

Six months of meetings, caucuses, hearings and interviews bore little progress. The Stark committee feared that if Congress failed to pass the proposed legislation then, it might be tabled for decades — a frustrating setback against the progress and future prosperity of agriculture.

This is how narrow the gap was between success and failure:

While the passage of the Plant Patent bill was being debated on the floor of the House, Congressman Fiorello La Guardia was informed of protests to the bill. La Guardia blocked action on it in response to the protests, based on an unsubstantiated comment by a staffer. (The comment was later determined to be inaccurate.)

Congressman Fred S. Purnell of Indiana, the bill's sponsor, pointedly walked over and asked La Guardia what he thought of Luther Burbank. La Guardia replied, "I think he is one of the greatest Americans who ever lived." Purnell, anticipating the answer in advance, replied, "I want to read a letter written by Luther Burbank just before his death."

To Paul C. Stark, Chairman, National Committee for Plant Patents:

"I have been for years in correspondence with leading breeders, nurserymen, and Federal officials and I despair of anything being done at present to secure to the plant breeder any adequate returns for his enormous outlays of energy and money. A man can patent a mouse trap or copyright a nasty song, but if he gives to the world a new fruit that will add millions to the value of earth's annual harvests he will be fortunate if he is rewarded by so much as having his name connected with the result. Though the surface of plant experimentation has thus far been only scratched and there is so much immeasurably important work waiting to be done in this line I would hesitate to advise a young man, no matter how gifted or devoted, to adopt plant breeding as a life work until America takes some action to protect his unquestioned rights to some benefit from his achievements.

Sincerely,
Luther Burbank"

Congressman La Guardia immediately requested the floor and said, "I withdraw my objection to this bill and move its adoption."

The bill passed the House of Representatives and was sent to the U. S. Senate, where it passed a short time later. Under this law, the plant breeder or the discoverer of an asexually-reproduced, new and distinctly different tree or plant could secure a patent for a period of 17 years. After that period of protection, the patent expired and anyone could then grow and sell that variety.

The impact of the law was tremendous, perhaps worthy of the hyperbole that appeared in the *Saint Paris (OK) News-Dispatch*:

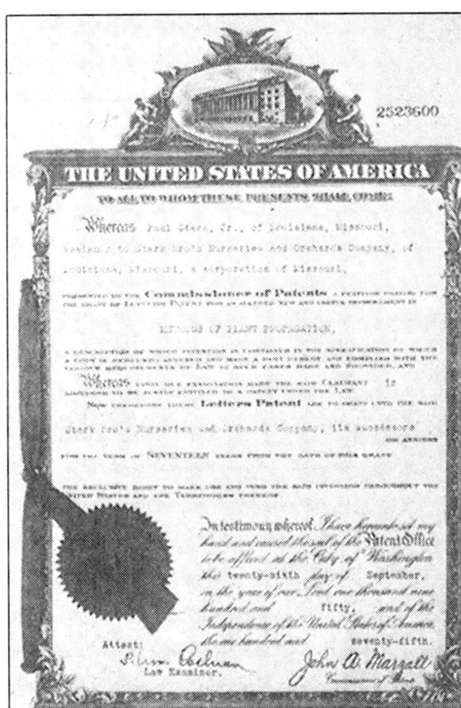

Stark Bro's was awarded the first-ever patent for a fruit tree, a great milestone.

"If the great horticulturists of the past ... could now come back for a day, they would gasp with astonishment when they learned that the strong arm of the law was now the protector of the toiler in the field of Fruit Tree Improvement ... a new day has dawned in the World's Fruit Tree History. Luther Burbank, Master Mind of all times in things horticultural, looked forward to this day, begged for this day ..."

The first patented fruit tree. Hal-Berta® Giant™ was the world's largest yellow-fleshed freestone peach. Below, the Starks celebrate the victory.

The first fruit tree patent — USPP7, for the Hal-Berta® Giant Peach™ — was granted on February 16, 1932. One can only imagine the success of this peach, had it not been the middle of the Great Depression.

Edgar worked with the banks and with Stark's finance department over those tough down years, helping to guide the business back to a sound and profitable operation.

Over the next few decades, Stark obtained many patents. They only have one that has not expired as of this writing: the Stark GrandGala™ Apple, 'Caitlin' Cultivar. In all, the company has held approximately 140 patents. Without them, there would likely be no Stark Bro's Nurseries & Orchards Co. today.

The 1970 Plant Variety Protection Act (PVPA)

The 1930 PPA legislation was broadened in 1970 with the Plant Variety Protection Act. It meant that seeds and propagated tubers could also be protected by utility patents, the same kind of patent used for a flea collar or a car part.

To be offered patent protection, a seed variety must not have been sold in the U. S. for longer than a year, or elsewhere for longer than four years. The variety must also reproduce reliably and be *distinct.*

"Distinctiveness" has always been a tussle, especially because DNA fingerprinting can now be used to unlock a plant's genetic code, some of which is just "junk," not expressing any trait.

Patents are valid for up to 25 years, after which anyone can propagate the plant for sale or otherwise. Someone could even then produce hybrid seeds, produced by crossing specific parents, because patents, available for anyone to see, spell out exactly how a product is made.

Trademarks vs. Patents

A patented plant is one thing; a trademark name is another. Patents have a limited life; trademarks can be renewed indefinitely. A plant might not have a patent, but it can be trademarked, a designation which is assigned to a company or an individual who could assign that name to any of their plants, even different ones. A patented variety also could be marketed and sold under multiple trademarks.

Stark Bro's also propagated several patented ornamentals, which were also protected.

13

The Record-Bearing Strains

The Victory Garden Program

Harry Guengerich, The Father of Virus-Indexing

13 | POWER CHANGES HANDS

1932-1953

The New Leadership

When Edgar Stark died in 1935, his son, Clay Hamilton Stark, succeeded him as company president. Lloyd Sr., who had been VP and General Manager, became Chairman and National Sales Director, and Paul Sr. continued as VP of Research and Production.

Edwin, Thomas and Lawrence, sons of then-Senator Eugene Stark, became VP, Secretary and Treasurer, respectively. Stickney, son of Lloyd Sr., became Assistant Secretary before entering the Navy in World War II.

Like his father, Clay H. was interested in finance and management. He was largely responsible for working out the budget plan under which the company operated for decades. Clay was also in charge of purchasing, real estate and legal matters. He traveled extensively around the world, always looking for new fruits, plants and production ideas.

Clay Hamilton Stark

The Starks always maintained that it was the patriotic duty of citizens to serve their country, and so Lloyd ventured into politics after his tenure with the family business. He already had a long record of public service, having been, among other things, chairman of the Citizens Road Bond Committee, which successfully campaigned for the $75,000,000 road bond issue in 1928. His friends urged him to run for Governor, which he did in 1936. He was elected by a large majority.

Lloyd Crow Stark

Governor Stark's administration is still remembered by Missourians for its integrity, and his personal reputation for putting the interests of the state above those of any faction or political group brought him national attention.

After his term ended in 1941, Gov. Stark returned to the nursery. He stayed on into his 80s, serving as Chairman.

Paul Clarence Stark Sr.

The Record-Bearing Strains

While Lloyd was busy being governor, members of his family were innovating. Fighting frost damage in budding apple trees had always been a challenge in a state that has extreme spring temperature fluctuations. Paul Stark Sr. had tinkered for years with ideas to combat the problem, and finally landed upon a cheap, workable solution: fight ice with ice.

When frost threatened to nip the apple buds in the spring of 1938, crews of men went into the orchards at night and sprayed the trees with water, forming a protective ice casing over the buds.

Spraying with water during a freeze can actually save the fruit

"We are not ready to recommend the practice to orchardists as yet," Stark said at the time, "but we have learned that perhaps as many as 50 percent of the buds have been saved by the ice film. We propose to continue our experiments." Mr. Stark explained moisture from the ice when it melts also helps to draw the frost out of the immature blossom. The practice is in universal use today.

The Starks also developed the famous "Record-Bearing" strains during this time. Research showed that certain trees of the same variety, grown under similar conditions would sometimes grow bigger crops of better fruit. This led to the practice of searching out the tree of each variety bearing the biggest crops and most superior fruit year after year.

Don Stinson reporting on a record-bearing tree

New trees were propagated "direct-in-line" from the selected record-bearing parent trees. This eliminated the possibility of degenerate strains creeping into the orchards.

The Victory Garden Program

During the food shortage of early 1945, President Harry S. Truman named Paul Stark Sr. as Director of Home Food Supply for the War Food Administration.

He was tasked with coordinating government agencies and private organizations to assist and encourage home food production and preservation.

The importance of this program is reflected by President Harry S. Truman's statement of June 2, 1945:

> *"In this fourth year of the war, the need for every ounce of food the American people can produce and preserve is greater than ever before. We Americans must do our part to help ... by growing a Victory Garden — whether it be in the backyard, in a community or company employee plot, or on the farm. With millions of American men and women dedicated to this task, our food will make a real contribution to the final victory and the peace."*

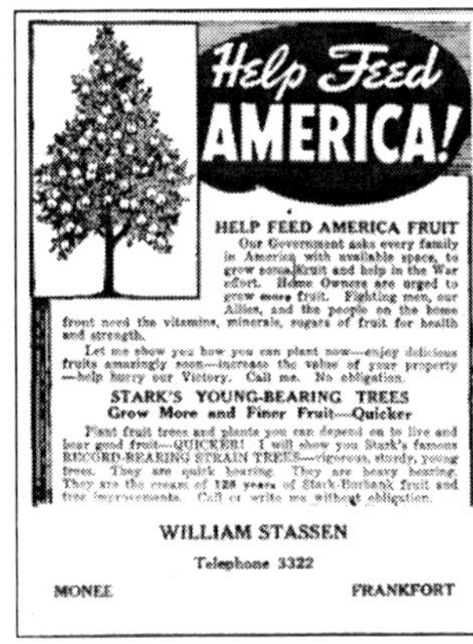

WWII-era ad for record-bearing strain trees

Paul Sr. served as president of the National Victory Garden Institute, an organization of forward-thinking business and civic leaders. It was this group that originally awakened Washington to the need for home production of vegetables and fruits.

They mobilized the nation's industry leaders for the Victory Garden program, who devised marketing programs and public service announcements to inform the public. Vegetable plots and fruit trees sprung up across the nation, and the program became a raging success. In 1944, roughly 43 percent of the fresh vegetables consumed by the public were grown in Victory Gardens, according to a USDA survey.

Paul Sr. produced this guide.

The Not-So-Maximum Security POW Camp

During World War II, there were more than 1.5 million POWs housed in camps all over the U. S. — one of them right down the road from Stark Bro's. The mostly Italian and German prisoners were held in a converted National Youth Administration facility at the corner of 14th and North Carolina from 1943-1946. Lax security allowed many of them to work as cheap labor for Stark Bro's during that down time.

Harry Guengerich, The Father of Virus Indexing

By all reckoning, Harry Guengerich was the second coming of Luther Burbank at Stark Bro's. He was hired in 1945, after nearly 30 years of previous experience working with Mexican fruit growers (advancing their growing methods by decades), and on variety improvement, development of interstem dwarf trees and virus certification.

Guengerich grew up with a backyard orchard, as did many children of that era, and developed his interests there. His college years were spent as a horticulture major at the University of Missouri, and when he graduated, he immediately went to work at a commercial orchard. Later, he turned to research, his real forte.

After landing at Stark Bro's, Guengerich began analyzing new varieties of fruit sent in by orchardists around the country and tested their viability through propagation. He would get scions and graft them onto established trees in a test orchard; if the fruit demonstrated superior qualities — like flavor, color, disease-resistance or keeping quality — Stark Bro's would buy the rights. (Most new varieties arose out of bud-sport mutations, but some came through artificial pollination. Guengerich never put much stock in that method, describing it as a "numbers game," with one tree in thousands possibly bearing better fruit.)

The chance seedling would also be checked for viruses. Harry discovered that many trees are symptomless carriers, a previously unknown fact.

"In order to determine whether a tree is virus-free, you implant a small piece of tissue from the suspect into an extremely sensitive variety," Harry said, describing the process of virus indexing. "It is very important for nurseries to work with virus-free stock. Viruses are specific, causing slow deterioration of one variety. In time, the infected trees lose vigor and eventually die."

If a chance seedling is infected, a scion is grafted onto nursery stock and then the potted tree is heat-treated. Often, the heat treatments will kill or slow down the virus enough that the tip of the grafted tree will grow faster than the virus. This tip is then taken as a scion and grafted into nursery stock and propagated.

Hopefully, this new plant will grow virus-free and produce a healthy specimen of a new variety of apple, which can be re-propagated and introduced to the market.

Harry, checking a quarantined heat-treated plant in 1956

Today, Stark Bro's research department sometimes takes advantage of serological and molecular techniques, but for the most part, most virus-testing is still done by indexing.

During his 30-plus years as Director of Research, his accomplishments and discoveries were many. Harry published articles on virus control and identification in the journal of the American Phyopathological Society and was the foremost authority in his field. He was known worldwide for his contributions to the fruit industry and often consulted leading horticulturists from other countries.

It was gratifying for the family that such experts came to Stark Bro's for help and advice; it meant they had hired very well. Harry was openly and respectfully referred to by his co-workers as "second to none" in tree research.

Guengerich's ultimate task was to please the fruit growers, the wholesalers, the retail sellers and the apple-eating consumers — a tall order. He also determined whether a new variety would be a commercial hit, and whether the tree could be propagated virus-free. The knowledge, time and patience required to successfully execute all of those elements is nothing short of astonishing — but Harry did it with the regularity of an atomic clock.

He provided personal advice on tree care for Stark customers into his 80s. He passed away in 1981 at the age of 86.

14

Out with the Old

Stark Bro's Dwarfs the Competition

1966: The 150th Anniversary

All in the Family

The Experimental Period

From Pencil to Processor

—1953-1979—

14 | FROM MULES TO MECHANIZATION

1953-1979

Clay Hamilton Stark died in 1953, while still serving as the company president. He was succeeded by Edwin Jackson Stark, his cousin. Edwin would oversee the biggest development in appledom since the discoveries of the Red and Golden Delicious apples.

1949 Officers of Stark Bro's Nurseries & Orchards Co. Seated: Paul Stark Jr. Left to right: John Stark Logan, Edwin Stark, Lloyd Stark, Clay Stark, Paul Stark Sr., John W. Stark and Lawrence Stark.

Not much is known about Edwin, but he graduated from the University of Missouri horticultural school and traveled the world each summer looking for new varieties, just as so many of his relatives had done.

He copyrighted several of the company's guides and booklets, but he was by all accounts, a quiet president and very well-liked by the employees. Inexplicably, he wrote a recurring Q&A column on orcharding for *Leghorn World*, the journal of the poultry-breeding community.

A Treat for Our Boys In Korea

In the early 1950s, large quantities of trees and plants were being shipped to every state in the Union, to Canada, Mexico and many other foreign countries. The Starks received letters from soldiers stationed in Korea during the war, those who were surprised and happy to find Delicious apples growing in abundance there.

One orchard, a gift from the Starks, was planted at a Presbyterian camp in Korea. During the Korean War, American troops on the march spotted its familiar red fruit. The men broke ranks and swarmed the trees, loading their pockets and knapsacks with apples. An embarrassed officer repaid the orchard owner with a requisition "for necessary fruit."

Out with the Old

Paul Stark Jr. (left), along with an unknown helper, plowing weeds in the '50s. Homer Reed's machine replaced the mules.

The 1950s were a time of great change for Stark Bro's, which could not be more crisply illustrated than in the following paragraphs.

This memo was issued in the early 1950s to the workers:

"On or about Nov. 20th, all mules are to leave the Louisiana ranch. Returns from the sale of these mules will be used as payment for a narrow shade-tree row tractor that will be necessitated by their departure."

At least some of the stubborn critters remained until 1962.

In stark contrast, high-tech was about to make its debut in the company offices. In the early 1950s, the Mail Room and old office were joined to accommodate the nursery's first automation effort: an IBM punch-card system spearheaded by John Logan. That would be the genesis of Stark's first customer database.

Marketing methods underwent some changes as well, and two new programs were introduced. In the 1950s, Stark Bro's (and other nurseries) sold a package of "landscape plans" along with the bare root plants listed in the Stark mail-order catalog.

A couple of Edwin's landscape books

"City Sales" customers bought the "plan" first from a glossy color catalog, and then the bare-root plants arrived one to 10 months later for the homeowner to plant.

Retail customers could now buy mail-order items from the door-to-door salesmen. At the height of this program, there were 15,000 "spare-timers." Many were retired farmers, merchants, schoolteachers and horticulturists who worked on a fat 35% commission and mostly sold to themselves to get the discount.

Relatively few fruit trees were sold under either program, but they were supported for another 20 years or so.

Stark Bro's Dwarfs the Competition

Original spur-type semi-dwarf tree of Starkrimson® Delicious, Bisbee cultivar, in its burglar-proof cage.

Since about 1923, cultivar improvement had been focused on bud sports (natural mutations). It was now Paul Stark Jr. who had responsibility for seeking out new bud sports and recommending them to the Stark Bro's Board for introduction.

In his annual summer sojourn, Paul found one of the most important discoveries in Stark history: the spur-type, semi-dwarf Starkrimson® Red Delicious ('Bisbee' cultivar). He wired the company his recommendation to purchase it from Roy Bisbee, and paid $25,000 for it on the spot.

It proved to be the greatest red apple they had ever introduced; it can be picked with 100 percent extra-fancy color earlier, and stays crisp and juicy longer. It was by far the most expensive tree purchase to date.

There had been spur-type semi-dwarf Delicious apple trees before, but their fruit did not have the same quality. The Bisbee tree did. It produced almost twice as many apples per acre as a standard-size tree, with earlier and more consistent color, a nice, long shape and better keeping ability.

Not only was the fruit amazing, but so were the possibilities for the way it could be grown. The spur-type semi-dwarf grows to only two-thirds the height of the standard size, can be planted much more densely and bears much younger than standard trees, which are 25 to 30 feet tall.

Standard tree (top) vs. Spur-Type Semi-Dwarf

But the real beauty of a spur-type tree is the way it bears fruit. A standard tree will produce its fruit on the outer ends of the limbs, far from the trunk. The spur-type tree bears clusters of fruit all the

Four Stark Generations. (Left) Paul Sr., Lloyd Jr., James Stark, Edwin J. Stark (with Gov. Robert Holmes, Oregon and Sen. Harry Byrd, Jr., Virginia) and Paul Jr. view original semi-dwarf Starkrimson® at its dedication.

way up and down the limbs inside the tree, and lots of it. Growers were ecstatic, reporting an average of six bushels per tree. This was unheard of before the spur-type was discovered.

Semi-dwarf trees, planted at high densities, are also easier to harvest and prune, saving thousands in labor costs. Competing nurseries jumped on this two-for-one miracle, fighting to maintain their Stark exclusives and simultaneously come up with something better of their own.

Paul Stark Jr., the Snyders at C&O Nursery (Washington), the Van Well family (Washington) and the Heusers with Hilltop (Michigan) were all major players in the period from the early '60s through the late '80s. Stark Bro's has developed many other outstanding spur-type varieties since that time. They are now growing Stark Golden Delicious, Red Rome Beauty, 'Winesap', 'Arkansas Black', 'UltraMac'™ and even a 'Montmorency Pie Cherry.'

The Starks' next great bud-sport discovery was in the orchard of Elon Gilbert, near Yakima, Washington. For the Starkspur® Golden Delicious, the company paid a whopping $51,000, the highest price in history for a fruit tree. It was a bargain, considering it would produce millions of trees that would sell themselves.

Starkspur® Golden Delicious, perhaps Stark Bro's greatest variety

Now growers had these marvelous spur-type semi-dwarfs in both red *and* yellow. Home orchardists with small properties were rejoicing.

"The spur-type tree has more to offer planters than anything we have ever seen," said Grady Auvil, the famous orchardist who helped pioneer Granny Smith, Gala and Fuji apples in the Northwest.

Times were very, very good for Stark Bro's, thanks to Paul Stark Jr.'s discoveries.

THE FOUR-PART INTERSTEM TREE

Stark Bro's took the dwarfing process one step further, with their "invention" of the four-part interstem tree, the brainchild of Research Director Harry Guengerich. His challenge: to test and improve the dwarf apple tree, or as it would be called in the catalogs, "the four-part interstem tree."

Dwarfing rootstocks have the tendency of tipping, so Harry started experimenting with standard roots (that have good anchorage) and a dwarfing piece in the middle of the trunk. This historical assignment earned Stark Bro's a patent for the process.

Well into the 1970s, Paul Jr. was immersed in working with interstems and high-density orchards. In a research paper that drew wide acclaim, Paul Jr. predicted that dwarf interstem trees "will be the tree of the future, even over the Malling rootstocks."

The idea was that smaller, more efficiently-producing interstem trees got the grower into production more quickly because they bore fruit considerably faster than standard-sized trees. The interstem trees were also adaptable to a broad variety of soil and climate conditions. The other bonus was that much of the labor could be done from ground level.

Budding techniques and rootstocks have improved greatly since then, so there is no longer a need for the interstem tree. One thing is for sure: If Harry Guengerich were still here, he would have been the one to discover its successor.

In 1961, Ginny Andrews graduated from Louisiana High School. Three days later, she started working at Stark Bro's in the City Sales office typing pool. Her pay was 45 cents an hour.

"(President) Edwin Stark used to tease me and say I looked like Jacqueline Kennedy," said Ginny, who retired after 48 years in 2009. "We were all very close then, it was a wonderful place to be." She still works part-time in the Garden Center during the busy spring season.

Ginny was there when John Stark Logan became president upon the passing of Edwin, a humble man who chose not to have a funeral. John had been with the company for a long while already, and was up to the task.

"We had an ongoing mutual respect for one another," said Jim Law, former VP of Production. "John carried both the responsibility of President and Chairman, and also had to contend with different factions of the families in keeping them informed and 'happy.' I feel very privileged to have known and worked with him."

In 1964, Stark Bro's gave a Ford Fairlane to its best salesman. Ginny Andrews is in the driver's seat.

Dozens of other people interviewed for this book expressed similar sentiments for the man known as "Quiet One."

In 1964, John Logan's first year — and 50 years after the discovery of the Golden Delicious — the entire State of West Virginia paid homage to it. The Governor of West Virginia, U. S. Senators and other dignitaries took part in the events. Representatives of the USDA and even ambassadors from foreign countries came to salute this one single apple.

Edwin Stark (left) and Edwin Gould

Edwin Gould, Superintendent of the University of West Virginia Experimental Fruit Farm, and one of the foremost authorities on bud sports, was a speaker.

"The discovery of the Golden Delicious," he said, "was one of the most important developments in apple history."

John Logan was presented a figurative gift by the First Lady in 1965, when she launched her "Keep America Beautiful" campaign. Stark Bro's has always donated trees to improve the scenery, something that pleased them greatly — but this time there was a national figure telling everyone to go out and *buy* trees.

The "Keep America Beautiful" campaign caught on; in 1965, Stark shipped out nearly 2,500,000 fruit trees alone from its five-acre packing plant. The whole country was undergoing a makeover, much to the delight of the busy Stark accountants.

1966: The 150th Anniversary of Stark Bro's

Only a small handful belong to the elite group of businesses that have continuously operated for 150 years. The best testament to Stark Bro's survival is to walk into a grocery store and see the Red and Golden Delicious apples that are still ubiquitous. Such an accomplishment was worth a splashy celebration.

President John Logan served as emcee for the regal affair, which included a redux of the 1893 International New Fruit Show (which had produced the original Delicious apple), a parade of prominent speakers and the ceremonious unveiling of a Luther Burbank portrait at the closing.

Employee service earned a highly visible spot on the program as well. Everyone knew that without the "other" Stark family, the company would have never come so far.

The Employee Service Awards ceremony

Paul Sr., Paul Jr. and former governor Lloyd Crow Stark all made introductions. Seeing them all on the stage together was kind of like a Beatles reunion; they weren't all there, but it was plenty good enough.

"I had the privilege of working with Paul Sr.," said Jim Law, former VP of Production for many years. He lived and worked in the shadow of his brother Lloyd, but he was a remarkable man in his own right." At the 100th anniversary in 1916, President Edgar Stark had said:

> *"One hundred years ago the trees of old Judge James Stark became famous far and near, all over this section. For decades before the Civil War, planters made journeys of days and weeks by ox carts or wagon train to get 'those sure-growth trees Judge Stark raises up in Pike County'.*
>
> *"Years went by and the old Judge passed on. But the ideals he cherished lived ... so it has been that ever since 1816, the name of Stark Bro's has stood for all that grew best in nature for the good of mankind. This is our heritage, and this is our task, our duty: the preservation of this good reputation, the continuation of the public faith in the integrity of every promise, every claim, every tree that Stark Bro's offers."*

At the 150th celebration, John Stark Logan (Edgar's grandson by marriage) said, "... Except for the fact the company has been doing over eight times as much business as it did in 1916, nothing has changed in [Edgar's] policy."

At the end of the program, Luther's portrait was revealed. To this day, he is a sort of spiritual presence there; when his name is mentioned, it is almost said in a reverent way. Everyone there knows they are tasked with carrying on his work, as he wished them to do in his dying days.

All in the Family

In 1966, the business was in the hands of five Starks: three younger men and two "elder statesmen." John Logan and Paul Sr. (the latter) administered the business and instilled the team spirit that had been so successful in previous generations. They oversaw finances, operations and the office.

Paul Jr.'s main duties were in production and research, but he was about to get married (his third) and move to Washington state for about 10 years. Jim Law, who had been with Stark Bro's since 1965 as Paul Jr.'s assistant, and then as Production Manager, was named VP of Production in 1967 when his boss up and moved away.

Paul Stark Jr. and his famed dimples

Paul Jr. was a talented horticulturist and a charming, charismatic individual, but he had his foibles: financial problems, failed marriages (he would have four wives before he passed away) and a serious drinking problem, especially in later years.

Lloyd Stickney Stark Jr. was the sixth generation in the firm, and supervised City Sales. Another personable and somewhat charming fellow, but by all accounts, he was a wanderer and not exactly focused on his work.

Despite their contributions, these two were ultimately a burden to both the nursery and to John Logan, according to a close relative. Their immature behavior and continual disruption of the business led John to begin looking outside the family for more and more non-family managers in key positions.

John credited the company attorney from St. Louis, Bill Biggs, with "helping keep the peace" amongst the family members — many "type-A" personalities, and some with chronic money needs. John had severe stomach ulcers in the early '60s, and it is not hard to understand why.

The Board of Directors in 1966 (left to right): Paul Sr., John Logan, Lloyd Sr. Paul Jr. and Lloyd Jr.

Around this time, 14-year-old Walter Cyrus Logan II was becoming aware of the business.

"Even at that age, I was kind of rebelling against automatically being inducted into the family business. People said, 'you'll go to college, but come back to the business.' At the time I was non-committal, just to avoid the expectations."

Over in the typing pool, it was Rose Marie Buchholz's first day at work. She used a manual typewriter and onion-skin paper layered with sheets of messy carbon. The auditors would be reviewing what she typed, so no mistakes were allowed. There were no paper clips or staplers; to keep papers together, she used straight pins.

The "IBM Department"

In another office not far away, "the IBM Department" was trying to figure out how to create a database. That was in 1966.

That year, when Stark celebrated its 150th anniversary, half its sales volume came through two channels: door-to-door sales and City Sales' home landscape plans. Mail-order and commercial sales made up the rest. The ability to be nimble enough to flip those revenue streams would see the company through yet another coming downtime.

The Experimental Period

Back then, Stark management had cozied up to the idea that decentralizing operations would give them a hedge against the risk of bad weather or untimely growing seasons in any one part of the country.

In 1966, Stark Bro's had branch locations in Farland, California; Marionville,

Stark Bro's Logos Through the Years

1893

1950

1976

2016

Missouri; Farmington, Arkansas; Checotah, Oklahoma; Porum, Oklahoma and Atlas, Illinois, which is the only branch that remains today.

Around the same time, John Logan tried "an experiment:" two franchise Garden Centers (in Texas and Lamar, Missouri). They carried everything from grass and flowers to large shade trees, and were run by area managers from City Sales. These fellows were not well-versed in retail garden center operations; they were hiring managers.

Expectedly, the quality of the merchandise suffered because it was handled by salesmen, not nurserymen. Customers were disappointed, sales declined and worse, Stark's sterling reputation was being tarnished. Poor oversight ran the franchise operations into the ground within months.

After the quick failure of the franchises, Stark Fruit Tree Centers began selling branded, bare-root fruit trees and plants to "authorized dealers." These dealers were established garden centers in major cities in the central and eastern states.

"Stark Trees," with Stark picture tags, gave the products a distinct identity amidst the other generic offerings. Customers could now hop in the car and go look at Stark stock in person. By 1980, there were 1,200 dealers. The program remained successful into the 1980s, when the mega-garden centers took over.

The team also tried to shake things up by having a new logo designed. It was commonly referred to around the office as the "frying-pan-with-eggs" logo. The Stark Bear logo was thereafter used for the Commercial Division. That was 1970, and Clay Stark Logan, Walter's older brother, had just graduated from Westminster College. He joined the business as a trainee.

All Stark men now had executive titles. The Board ballooned from four during the era of black-

Young Walter picking a peach from a mature dwarf tree not much bigger than he was.

One Jumbo apple, big enough to make a pie

sheep William P., to 28 of them in 1971. Said Clay, "at one time there were so many Assistant-that and VPs of This, it was kind of silly."

All of this brain power would contribute to the development of the next "big" thing: the dwarf fruit tree, just eight to ten feet tall. There was even a double-dwarf. Now customers could buy fruit trees that grew in pots, which opened up the lucrative small-space gardeners' market.

The oxymoronic dwarf Jumbo was a fun offshoot, an apple with the Delicious pedigree, but decidedly un-dwarf in size. From an eight-foot tree came apples weighing two pounds or more, with a circumference of 15-20 inches each. This surreal fruit was the result of eight years of research by the great Harry Guengerich, Director of Research.

In 1992, Dan Fitzkee of Lancaster, PA grew a record-breaking Jumbo apple that weighted 2 pounds, 11.2 ounces and was 19 inches around, about the same size as a cantaloupe.

From Pencil to Processor

In the early 1970s, there were only four or five outside phone lines because of limitations by the telephone company. All internal and outgoing calls went through an old-fashioned switchboard — the kind with an operator who manually connected calls by moving cords from one jack to another. Only 30 or so people out of 300+ employees even had a desk phone.

"Toll-free" wasn't yet in the lexicon. The opinion of many in customer service was that called-in questions were a "nuisance," and that even order calls took time away from their real job: answering letters.

The office phone system was finally upgraded in the mid-'70s in conjunction with the creation of a formal inbound telemarketing department, created under Ginny Andrews' leadership. As a result, phone business increased dramatically.

Another sea-change was on the horizon, just as it was when Homer Reed, Carl Ince and Ernie Elder were inventing ways to mechanize processes in the

The legendary Carl Ince, Superintendent (left) and Shop Foreman Ernie Elder (right)

fields and sheds: technology was on the horizon. For the first time, the ancient ledgers and handwritten forms were going to be done away with; Stark Bro's first mainframe computer — a $50,000 purchase sporting a "huge" 200MB hard drive — and a raft of new electric typewriters had arrived.

Clay always wanted an inventory system. Data used to be stored on an old Syquest backup disk, and was shuttled back and forth to people for the counts. Clay and John dreamed of a day when the numbers could just be entered on a computer and available to everybody at the same time. John Logan was prescient in believing that this function was the business core: the interface between sales, marketing and production.

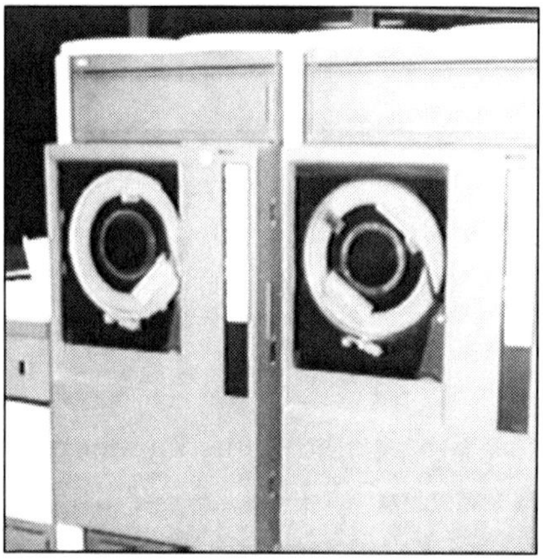

The behemoth IBM mainframe

John Logan had gone to IBM "school" back in the '60s to learn how these new mainframe computers were going to revolutionize the business world. He was instrumental in designing the templates for many of the initial mainframe reports. There was always a backlog of requests for new reports, changes to old reports, etc. and it got to point where the requests had to be "triaged."

This helped justify the later purchase of the TRS-80 computer, which debuted on the market in 1977 and cost $599.95, a small fortune in those days. Inventory remained a sort of hybrid system as late as the early '90s, with field records kept on books and cards, and the totals entered into the computer system.

The machines were coming.

—1953-1979—

The Stark Way

Erstwhile efforts to establish a formal "up the ladder" protocol in the Stark business were never put to paper, but there was an unspoken process that was well-known inside the family: as soon as they were old enough to know an apple tree from a pear tree, young Starks could go to work during summer vacations in the nurseries. By the time they went to college, they could be well-grounded in basic horticulture, if they wanted to learn.

Each department was headed by a Stark. In a practice that flew in the face of conventional wisdom, the young men in the family were given a free rein; as they came to grasp the executive responsibilities of the company, the young ones stepped up to the administrative tier of the company and the older men stepped back into the role of advisers and members of the Board of Directors, which met monthly to review policies, important developments and plans for the future.

Walter (left) and Clay Logan, in 1991. They were the last Starks to run the company.

Since Paul C. Stark Sr. attended the first horticulture school at Cornell at the turn of the century, almost all male Starks studied either agriculture or horticulture. Even those with business majors were practiced in the art of fruit-growing. It was just expected.

It was also expected that there would be a perpetual supply of Starks to keep the business going, because there always was. Oddly, there was never a formal succession plan. Clay had one son and Walter had three children, and other branches of the family had been prolific, but the seventh generation remained a question mark.

The Starks, unlike some other family "dynasties," never groomed their children or "pre-decided" that they would join the business. As it turns out, Clay and Walter's children have all taken other paths, and so the brothers are, quite literally, the last Starks to run the company.

15

The Fire

The Last of the Stark Nurserymen

The Gala 'Tenroy' Impostor

The Unionization Attempt

New Methods and a New Home

15 | FROM THE ASHES, A COUP

1974-1979

The Fire

Elmer Kidd was lost in his grafting rhythm early on February 20, 1974, when Loretta Fuhrman came in, sat down and said, "It's gone. 'N' Building is burning, it's gone."

The 1974 fire completely destroyed the "N" building. Suppliers pitched in quickly to help, and the season was saved.

A young man had backed one of the electric carts into a smudge pot, tipping it over and spilling oil onto the dirt floor.

One of the men threw water on it instead of smothering it with dirt. That pushed the fire to the old wood wall, where it caught onto the paint and papers plastered to it. The flames quickly spread to the false ceiling.

"They said you could see the fire all the way to Bowling Green," said Elmer Kidd. "We stayed there all night putting out hot-spots on the tar roof. A fumigator that used a toxic chemical was stored in that warehouse, and if it had exploded, there would have been many casualties."

Albert Bass stood there with a little hose, trying to keep the fire away from it," recalled Kidd. "It was an act of bravery on his part."

The fire incinerated more than a million ornamental plants, burning so fast and so hot that nothing was left but nails on the dirt floor. Luckily, no one was hurt. Appealing to benevolent and very responsive suppliers, Stark Bro's filled most of its spring orders, and with a huge sigh of relief, got through the season.

The fire caused a reassessment of the company's business direction and a shuffling of allocations. City Sales had been in decline for a while, and the big garden centers and landscape contractors were the final blow. Stark exited the City Sales landscape planning business within two years, and abandoned door-to-door sales the year after that. Being able to re-purpose the inventory was a help, because demand was enveloping the supply. These decisions were easy ones because although those two programs weren't bearing much fruit, so to speak, the mail-order catalog and commercial sides of the business were brisk.

Walter Logan was not there for the fire. He had graduated from Westminster College, but instead of joining the business, moved to Texas to work for Miller Brewing Co. He also went to "keep an eye on" his fiancée, who lived there.

"My dad was OK with that," said Walter. "Mom tried to steer me back, but Dad realized there would be value in me seeing how another company operates."

Walter and Cindy, who met on a blind date ("Cindy said she was the blind one," recalls Walter), were married in Texas in 1975. After that date, a sorority sister joked, "Cindy, you're gonna end up in Louisiana, Missouri, picking apples," to which Cindy replied, "No way in hell, I'm from Texas!" Within a year after their marriage, she and Walter would both be back in Louisiana.

The Last of the Stark Nurserymen

Walter did indeed join the business in 1976, albeit at a disadvantage. He had no interest in horticulture in high school, and got his college degree from the business school. For most Starks, the nursery supplied their first job.

"I think psychologically, I just didn't want to automatically work for the business," said Walter. "And I didn't do the summer job thing because of allergies. I grew up drinking Benadryl."

Clay (left), John (seated) and Walter (right)

During his first year, as are all Starks, Walter was a trainee in Atlas, Illinois, moving from department to department depending upon the season.

One of his more memorable moments was out in the fields that October.

"You have to force the trees into dormancy for the winter, so you strip the leaves off," he explained. "I had a pair of jersey gloves on, and a couple of old-timers were showing me the ropes. After a couple of days stripping leaves, my gloves wore out.

"The old timers, hoping to show me up, said, 'I've had these gloves for months and they haven't done that,' and I said, "Well, I guess that shows you who's doin' all the work out here."

The next day, Walter and another employee were assigned to move the outhouse, a memorable event of the less jovial type.

Business was good: as of 1976, about 60 percent of trees sold were retail, about 40 percent wholesale. The novel notion of a toll-free number was tested to see what kind of impact it would have on response rates and order size. The answer would be an emphatic "big," and beginning soon thereafter, the toll-free number was offered.

Workers grafting young trees in the field, 1978

Business may have been good, but Paul Jr. was not so good, and seemed to be on a downward slide. There were reports of money trouble; he was apparently unable to repay a number of loans collateralized by his Stark Bro's stock, which the banks sold to Paul Sr. Then, after about 10 years of living with his second wife Wilma in Washington state, they divorced and prodigal son Paul Jr. returned, asking for his old job back.

It wasn't long before he married Wife #3, Hadley. According to a close family member, the newlyweds were drinking heavily and the marriage only lasted a few years. Hadley died in 2009; there is no mention of Paul Jr. in her obituary.

The Gala 'Tenroy' Impostor

The 1970s fruit market was focused on producing apples that tasted good, because buyers were tiring of buying beautiful, red apples that did not. The first Gala apple tree, discovered in 1934, was one of many varieties that came from a cross between Golden Delicious and Kidd's Orange Red, planted in New Zealand by orchardist J. H. Kidd.

"We signed an agreement with the patent owner to be the exclusive licensed agent in North America in 1974," explained Clay Logan. "We marketed the product, and paid the patent owner a royalty."

Gala apples are red-and-yellow striped, and so the search ensued to find a red sport of this variety, which turned out to be the darker-skinned Royal Gala®. The Royal Gala 'Tenroy' cultivar has its own story.

This bud sport, discovered in a New Zealand orchard, is significantly darker red all over versus the original Gala. Because consumers had a strong preference for red varieties, the "redder Gala" was a tremendous find for Stark Bro's.

Royal Gala 'Tenroy' apples

In 1977, Paul Stark Jr. acquired the U. S. and European rights to Tenroy from Bill ten Hove in New Zealand. Trials were very successful, and Royal Gala was the next variety to take the commercial market by storm. This new variety produced a windfall of significant profits.

After the industry heard about the "new red Gala," other nurseries started rumors that they, too, had a significantly redder Gala. In 1956, when Starkrimson® Red Delicious was introduced, within a few years other nurseries started patenting very similar Red Delicious varieties. By the early 1980s, well over 250 patents had been awarded for darker Red Delicious cultivars. About 25 of them had identical traits to Stark Bro's Bisbee cultivar.

"This became a nurseryman's joke," said Clay Logan. "How can there be 250+ 'different' Red Delicious varieties with enough distinction to be awarded a plant patent? There may have been a few dozen, but nowhere near the 250+ patented."

Fearing a repeat of the Red Delicious patent stampede after the introduction of Tenroy, the company launched an aggressive campaign in the mid-1980s to challenge patent applications that were not truly "different." The first target: their major mid-west competitor, Hilltop Fruit Trees in Hartford Michigan.

Clay retained a patent attorney from St. Louis and expert witness Dr. Jules Janick, a distinguished professor of horticulture from Purdue University. Dr. Janick was prepared to testify in court if he could find no horticultural differences between other nurseries' "red Gala" and Tenroy.

Clay, his research director, attorney and Dr. Janick traveled to Hilltop Nursery, and they were shown the apple in question. With the attorney as a witness, Dr.

Janick compared the apple to Tenroy, and found it to be identical.

Thankfully, so did the Hilltop folks, and they graciously dropped their efforts to patent their "red Gala" in return for a license to grow and market Tenroy.

The same Stark "Red Gala Posse" paid visits to several nurseries on the West Coast, looking at their varieties and leaving copies of the Tenroy plant patent for them to compare against. Soon, Stark had nearly a dozen U. S. nurseries licensed to grow Tenroy. Clay also managed to get a number of smaller nurseries to pay royalties retroactively on trees they had sold without a license.

The math was easy. Stark paid ten Hove about 25 cents per tree and charged their own licensees one dollar. For at least six years, 250,000-500,000 Royal Gala trees per year were grown and sold by Stark and its licensees. The sum: princely.

Recalls Clay, "We also made settlements with a number of large orchardists who had grafted trees to Tenroy in their orchards, or had Tenroy trees grown for them by small, unscrupulous growers. There was one huge California orchardist ... we couldn't catch him, but I'm certain he grew thousands of trees illegally. This orchardist will remain unnamed, but I will say that you probably drink some of that company's wine on occasion."

Licensing and Growing "Clubs"

The popular Honeycrisp™ apple is a product of the University of Minnesota. Consumers loved it and were willing to pay extra for it, and now there's a race to create the next Honeycrisp.

There is a critical difference in the way these apples came to market: they were controlled by the creator. If you bought a Honeycrisp tree, the grower had to pay a royalty to the university because of the patent.

If you want a SweeTango® tree, or a lot of them, you have to get them from the Next Big Thing cooperative. The University of Minnesota exclusively licensed SweeTango to a single group of 45 growers — mostly in Washington, Michigan and New York. SweeTango has a patent, but it also has a trademark, and that means the royalties will never go away. Understandably, growers are not happy, but proponents say that quality and quantity control, and the ability to "brand" a fruit give growers the edge in the end.

The days of exclusive variety control by one nursery are gone now. The Next Big Thing and other "growing clubs" are increasing in number.

The Unionization Attempt

It is not widely known — even to this day — but the first Stark management heard of the union attempt was when Walter's best friend and his younger sister (a Stark employee) were out boating on the river one day in April. The sister innocently asked the boys, "So what do you think of the union organizing effort?"

Walter clearly recalls his response as, "WHAT?"

"She laid out all the details," said Walter, which he later shared with Clay and John. It wasn't 10 minutes before they were on the phone to their attorney.

"Handling the union thing was a good learning experience for me," remembers Clay. "It started due to the frustration of some employees with the managers." John gave full responsibility for the issue to Clay.

According to some, Clay's management style was strictly "top down" and autocratic, which stifled morale and productivity, and encouraged high turnover in younger employees. Others viewed him simply as a "no-nonsense" leader. In actuality, he had only ever supervised one person, and that was a 70-year-old woman who assisted him with inventory records. Many of the production managers at the time had the same type of management style.

Looking back, Clay wondered if his dad's attempt to disengage himself from the day-to-day operations was to cultivate "outsiders" to hold management positions. Some Stark family members were meddling in each other's areas of responsibility, which created a huge amount of confusion and disruption. John may have stepped back a little too far, fomenting employee discontent and some interest in unionizing.

The nearby Dundee cement plant employees (who were members of the Cement, Lime and Gypsum Workers union) and some Stark workers were drinking buddies, so it seemed logical that the Dundee men would do a little recruiting. A small group of Stark employees were receptive. This was a serious jolt to management, who were used to calm seas on the employee front.

From the beginning, there was a question as to whether Stark was agricultural or industrial; the union rules are different for each. Mediation with the National Labor Relations Board went tersely forward. In September, the NLRB determined that Stark Bro's was agricultural, and did not fall under its auspices. John and Clay Logan took a collective deep breath, then decided to do something tangible to improve things and prevent another unionization attempt.

Enter Dr. James Miller, a consultant and practitioner of the popular "Participative Management" (PM) paradigm. The program involves personality testing, green-light meetings and a host of other interactive exercises to improve employee-management relations.

Under Clay's direction, successful programs changed the leadership style of almost all company managers and encouraged a high degree of employee participation. The result was greatly improved morale and productivity, and a lot more respect for those first-line managers. Another outcome was to reclassify many people, changing them from hourly to salaried. After that transpired, there was no more union talk. Some employees called it a payoff; others said it was deserved. In any event, everyone was satisfied, perhaps with the exception of one.

"I was on the wrong side of it as a young worker," said Elmer Kidd. "If there's one regret I have here, it's taking that stand then." Since 1977, Elmer has become known worldwide for his horticultural expertise and is today, by any measure, Stark's most valuable employee. There are no hard feelings.

New safety rules were also put in place for the workers, including this one: *All employees of the Company are required to wear shoes while on the Company premises.*

"We used to go barefoot all the time out in the fields," remembers Barb Brewer. "Then this new rule came down, so we'd wear our shoes into the field, slip 'em off and throw 'em down the row. We didn't care about the glass, thorns or anything. It was just so nice when they'd fresh-plow and you could walk barefooted through it, so nice."

Elmer Kidd, out in the production fields

Important employee comings and goings characterized the year 1977. Louis "Lou" Brown (the current owner's father) started working at Stark Bro's. He was a University of Illinois graduate with a biology degree, and owner of Brown's Nursery in Griggsville, Illinois. Lou was one of the few people working at Stark Bro's who had any formal horticultural education. He brought a needed layer of propagation, landscaping and plant taxonomy expertise to the company.

Carl Ince was headed for retirement. From field worker to Plant Superintendent, Carl Ince's storied career spanned 55 years. Paul Stark Jr. once remarked, "Carl was probably the best budder the nursery has had. Very few are ever able to master the skill as quickly and as successfully as he."

Carl was a hard, enthusiastic worker, kind and compassionate. He had no formal training; nonetheless, he became an expert at identification. He was in charge of hundreds of acres of trees, charged with making sure that all budding and grafting for each variety was done on schedule. It always was.

New Methods and a New Home

Barb Brewer also remembers 1978, when she started in Inventory, keeping entries manually in a handwritten ledger.

"I had a book for each warehouse," she recalls. "A few years later, we got our first computer system in the inventory office, a big old thing. It was hard at first, because we used to have so many catalog numbers that we had used for years. Rose Marie [Buchholz] and I still remember them all."

The workers got a copy of the field book, went out and hand-wrote the field inventory. Then they would bring the books back to the office, for keying in. Now that inventory is done with an iPad, eliminating five steps in the process.

The new toll-free number was creating enormous call volume. To handle it, Clay and Walter installed a trailer outside of the old Victorian office building. Up to 25 employees at a time were in the trailer, taking orders over the phone.

In the midst of all of these changes, Clay found the time to get married. He and his wife Sally met on a blind date, just like Walter and his wife Cindy.

The fall of 1979 brought tremendous excitement to the office workers, because they were about to have a new 24,000 square-foot home. The new corporate offices, across from the Garden Center and down the road apiece, would give them the space and modern comforts they had been without for so long. The old Victorian office building, originally built as a home in case the business ever failed, was razed because the second floor had become structurally unsound.

"The architects kept me abreast of the new office building's progress," said

That "Alien Spinning Machine"

The first use of electronically-transmitted data was when Stark Bro's Virginia advertising agency (the one that designed the "frying pan with eggs" logo) installed an odd-looking machine in the office. It had a cylinder that spun around with the document clinging to it, and this machine would "read" it and send it to the recipient, who had the same kind of machine. The recipient could then print what had been "read." It was quite the rage in the office, which used it send ad proofs back to the agency. Back then, it was called a facsimile machine. Later it was called a fax machine, which is already obsolete.

Rose Marie Buchholz, a Production Department employee since 1969.

"Honestly, I would rather they had put the money into restoring the old office building to preserve it," she said. "I was the last person to walk into every room, envision all the years of progress, and the last to close the door. We had to move into the modern age."

And so they did. Dozens of operators were hired to handle the phones and whirring computers. Mail Order was sending out nearly 10 million pieces of mail a year, generating about three-quarters of the company's revenue, while Commercial and the Fruit Tree Centers sold three-quarters of the trees. The Garden Center had not been "born" yet.

By the end of 1979, the prime interest rate was at 15.5 percent. Stark's debt load was mostly capital loans, but the burgeoning sales seemed to justify the interest expense, so no one minded much. Things just rolled along.

Walter finished his year of training, and took over Clay's old role as Inventory Control Manager. That position worked hand-in-glove with Sales and Production to match up forecasts with how much nursery stock to grow.

The new office building

"There was quite a rift between sales and production, because sales could sell anything production could grow," recalls Walter. "Production was growing as fast as they could and still maintain quality. We felt it was better to sell all the inventory than set a match to it at the end of May."

Unlike a company that manufactures "widgets," Stark Bro's products take two to four years to produce, depending upon the fruit type and propagation method. Adding a third shift to keep up was not going to make the trees grow faster. Stark Bro's did buy some more acreage in Arkansas to pump up production — but the minute supply caught up with demand, the economy went south.

John Logan always made jokes, no matter what. "Every time a nursery builds a big new office building, they have tough financial times," he would say. How prophetic that joke would be, and it wouldn't be one bit funny.

16

The Recession Hits

A Steep Hill for a New President

"Ginny's" Garden Center

Stark Bro's Fulfillment Starts Up

The Garden Center Becomes a Profit Center

16 | THE BOTTOM FALLS OUT

1980-1991

The early '80s were lean times for virtually every nursery in the U S. Belt-tightening saw most of the old-time companies through, but at Stark Bro's, the worm was threatening to ruin the apple — again.

The Recession Hits

In the early 1980s, Stark's production fell from a high of 4 million trees to about half that. Sales went from more than $15 million in 1979 to only $9 million in four years flat. Because the trees have a two-year growing cycle, they could not slow down the production to compensate for the lack of sales.

The biggest contributors to the dramatic sales drop were the huge Mexican peso devaluation and a tax law change that disallowed orchard depreciation. Along with high domestic inflation and interest rates approaching 20 percent, orchardists couldn't buy land or fruit trees to plant on it.

Add to that some stubborn-headed Stark folks who held on to the interstem tree while the rest of the industry was moving to the Malling rootstock. In sum: the company suffered major losses and bank loans were required to keep it afloat.

As if that weren't enough, the big-box garden centers like Home Depot and Kmart began springing up. Fruit trees were available dirt-cheap because they were such an overproduced commodity. Potential Stark Bro's customers could find trees for $8; why would they pay Stark $15?

To make matters even *worse*, the trees that were sold at the big-box stores had the wrong rootstocks; the varieties were not true to the parent. People bought what was supposed to be a dwarf tree, and it grew over the top of the house. Fruit trees had also acquired a reputation as fussy, time-consuming and labor-intensive. Customer disenchantment with fruit trees was growing, and so they stopped buying fruit trees — period. Predictably, Stark Bro's mail-order business started falling off the cliff.

Stark had become a big fish in a pond that was drying up. Way back when, every nursery had fruit trees; it wasn't a specialty. But once the country's orchards were planted, demand diminished. Fruit trees bear for a long time, and no one needed any more. Stark Bro's forayed into ornamentals to take up the slack, but they were known for fruit trees, and couldn't get a foothold. Fierce competition was coming from the Yakima and Wenatchee River Valley nurseries, too.

Clay left Inventory in 1977 and became Assistant to the President, Stark's "point person" to evolve the culture of the business and eventually take over the company reins. Existing management and many of the employees entered the company workforce prior to or shortly after WWII, so it was apparent that this generation was nearing retirement.

Walter was still in Inventory in 1980, when he was asked if he'd be interested in moving to Mail Order as Assistant Sales Manager. He was, and he did. Numbers held more interest for him; it was also the pulse that John kept his finger on.

Sixth (and seventh generation) Starks: Walter (kneeling, left); his daughter Bethany: sons, Blake and Benjamin; Clay and his son John.

"I could see the writing on the wall that he was not going to be around forever," said Walter. Clay was clearly taking the administrative path so Walter felt it would behoove him to go the Mail Order route.

Walter's new supervisor, an experienced sales manager, left shortly after Walter moved over, leaving him in a lurch. Fortunately, there were some key people, in the form of Lita Eatock and Pennye Bradshaw, who stepped in to help.

Jack Alexander began work as Purchasing Manager around this time. He and Clay had gone to kindergarten together, and were life-long friends.

On the modernization front, a next-generation phone switch in the new office building gave operators 100 phone stations. While the shine of new technology was getting all the attention, something had to suffer, and it was unfortunately research. Stark stopped chasing new varieties as they once did, and Research went from five or six people to Elmer Kidd and his spare time.

Every nursery was scraping the bottom, and it was a rather unscrupulous period for the industry. Some would try and save money by replanting their dug-up trees and selling them in another year, hoping the economy would recover. A couple of those enterprises went under in the process. Stark Bro's never went this route, but instead ended up destroying close to 4 million trees they couldn't sell. It devastated John Logan, to the point of surrender. Clay hated it just as much.

"Dad was past retirement age, and he said, 'Why don't you take over,'" Clay recalled, choking back tears as he recounted the conversation.

John Stark Logan formally retired in 1984, saying to his oldest son, "You'll start with a fresh slate because you'll have gone through that downturn." And so Clay Stark Logan became the last Stark family member to run the company.

A Steep Hill for a New President

Clay became president of a drowning company that could barely come up for air. The pressure to save his family's 166-year-old legacy must have been crushing.

"That was a difficult time, we were in kind of a freefall during that period. We didn't really do anything. We just had to adjust, not knowing whether this was a permanent trend or not," said Clay.

The recession and resulting industry shakeout torpedoed plant sales in general, but hit fruit-tree sales in particular. Some of the downturn was predictable.

Said Clay: "My dad had a chart tracking Stark's sales and profits back to the early 1900s, and you can see a very clear up-and-down cycle about every 10 years, so by the end of the decade, you would wonder what was coming. As a young guy, I thought, 'Good management ought to be able to level that out.' How naïve. More false words were never spoken."

So Stark Bro's continued to emphasize the high quality of its trees, but the business sector languished, dragging everything down. Layoffs, production cuts and burn piles were all too frequent, and even those were only band-aids. Profits were still tied to mail-order fruit trees, which were tied to the whims of buyers.

The branch locations were another setback. The popular philosophy was to hedge against the weather risks by growing geographically — great during good times, but during slow times like the late '70s-early '80s, the costs of maintaining all the equipment and vehicles for every location was prohibitive. So all of the branches were closed except Louisiana and Atlas, resulting in a rash of management layoffs.

"Clay had to make the hard decisions when we had to contract the business, and I don't think he's quite the people person I was," observed Walter. "I was aware he wasn't that well-liked. He was the bad cop, I was the good cop."

Plant Managers' Meeting in Louisiana, Missouri

The stress of the job surely contributed to his style, but apparently not much empathy was forthcoming.

"[Clay] once told me that he felt he had to be more formal at the office," recalls Marian Kuntz, the company's HR & Finance Manager. "He thought it was required of a person in his position. As time went on, he became less formal. At one of our company dinners, we had a dunking booth. He went in, green pants, yellow polo, belt and all ... I'm not sure he even took off his shoes!"

Budget cuts and layoffs became the norm, so Clay needed to stay strong. In a savage hit, the Stark's long-time bank decided they didn't want Stark as a customer anymore. So Clay switched to a lender in Rhode Island. It was seized by the FDIC in 1988, which forced a search for yet another lender.

The very expensive TRS-80

Said Clay: "I remember talking to a group of bankers, pitching to get a loan. One of them said: 'Let me get this straight. You want me to loan you money, with trees as collateral?' I said, yeah, 'that's right.' And he said, 'OK, go ahead.' We still have a strong relationship with [United] bank. They saved us."

The company did purchase that extravagant TRS-80 desktop computer

— certainly a stretch during such tight times — but Clay believed it would pay for itself in short order, and it did. Mail Order used it to analyze mailing tests and create various "what-if" models.

"Ginny's" Garden Center

There was one bright spot: Ginny Andrews, the telephone sales manager, was approached by her boss and asked if she'd be interested in the notion of running Lou Brown's "Garden Center In A Trailer" concept.

Ginny, who back in 1977 had written down her dream to run a garden center someday, got her wish. She could hardly contain her excitement, talking a mile-a-minute, as she is prone to do. It was a pin-dot of light in the midst of some very dark, low-hanging clouds.

Ginny Andrews and "her" Garden Center beginnings in 1983

For a while, there had been an issue with visitors continuously interrupting the daily routine of the field and grounds workers wanting to buy trees and ask questions. Lou Brown had an idea: put a trailer over across the railroad tracks and start selling hard goods down there. That was the motive for the Garden Center: to keep people from pestering the workers.

Now Ginny had taken over. It would be a few years before the Garden Center moved to its permanent home in the old Luther Burbank Building, but she needed time to learn about the trees and the chemicals, and she did. Ginny passed the Certified Nurseryman Exam, one of the first women in Missouri to do so.

Up in the offices, Walter finally took over as Mail Order Sales Manager, where he remained for the rest of his career. Jack Alexander, Clay's old friend, was given a promotion to the newly-created position of VP of Operations.

The recession pushed the company to find greater efficiencies in its operations, putting Walter at the forefront of finding solutions for the Mail Order division. He frequently tested catalog copy and photos against control versions, identifying combinations that brought in more revenue. Stats were analyzed, and databases were cleaned up; regular customers who bought a lot had their names rented to other companies. The Fruit Tree Centers began selling potted trees as well as bare-root, which also added to the bottom line.

It became clear that upselling and other telemarketing techniques made toll-free order lines a paying proposition, and so it was rolled out and heavily promoted. Not only were more and more orders being taken from toll-free lines, but customer service calls (versus letters) dramatically increased. At one point there were 50 incoming phone lines, a sharp contrast to the four or five lines they were limited to in 1972.

A small portion of the Call Center

Later, there was an outbound telemarketing program. It was used primarily as a follow-up for catalog requesters who had provided a phone number. If they hadn't ordered within a certain number of weeks, an operator would call, asking if they had any questions and give them an incentive to become a customer. That was a success and was permanently implemented.

Some things worked; others didn't. Stark Bro's had gone shopping for mail-order companies with established names and potentially lucrative customer lists. In 1984, Clay bought Inter-State Nurseries of Hamburg, Iowa, one of the nation's top mail-order rose retailers. Four years later, he bought Kelly Nurseries of Dansville, New York. Both had filed for bankruptcy protection before Stark came to them. Clay and Walter hoped to assume their catalogs and apply the Stark touch to their mailing lists.

Another catalog, Bakker of Holland, was a joint venture with a Dutch nursery. This bulb-selling program was successful, but a declining dollar forced Stark to drop it in 1990. The currency exchange rate for the guilder tripled in Holland's favor, and a huge spike in postal rates "put the nail in the coffin" of that deal, according to Clay. The Inter-State and Kelly catalogs also flopped.

Stark Bro's tried publishing a separate catalog featuring ornamentals and perennials (mostly purchased from other growers). That catalog was created to appeal to a younger, more suburban audience versus the older, rural customers, who had been the predominant profile for decades.

The program lasted only two years. "We couldn't make it go," said Clay. "The [Stark] name was so linked to fruit."

Stark Bro's Fulfillment Services Starts Up

Along about 1984 there was a real push for Stark Bro's to transform into a more flexible/less vulnerable firm. The management team revised their target markets, shook up their product mix and packaging, and added a new profit center: Fulfillment Services.

Many core Stark activities were seasonal, of course. Managers would hire 100-200 people and ramp up, then lay workers off at season's end. The train/hire/layoff cycle should have been train/hire/keep. Buildings and warehouses should not be sitting empty in the off-season. So Stark Bro's Fulfillment Services was formed, and allocated 50 employees to work with direct-marketing companies to fill their orders. Jack Alexander was put in charge. It was happy news for seasonal workers who desperately wanted a year-'round paycheck.

Stark Bro's got a taste of this business earlier on through a sizable contract with Armstrong Nurseries, so they knew the ropes and could offer their clients a range of services, including telemarketing, order entry, order processing, product warehousing, order shipping and customer service. The division also had contracts for procurement and inventory management. Those 50 employees filled over 1 million orders and took over 1 million phone calls, qualifying the start-up division as a solid success.

One of its first fulfillment clients was Foster & Gallagher, who needed someone to store the product and ship orders for their hybrid tea rose catalog.

The Garden Center Becomes a Profit Center

In 1986, the old "seed house," also known as the Luther Burbank Building, was converted to a permanent Garden Center. John Stoeckley and Ginny Andrews worked together on the design and layout, and had a big grand opening. It is one of the most comprehensively stocked garden centers in the state.

John Stoeckley was recruited in 1985 as VP of Marketing. This move was at

The Stark Bro's Garden Center

the bank's insistence; they wanted management improvements to address weaknesses in a still-shaky company.

In addition to redesigning the catalog and getting the Garden Center off the ground, John made many significant contributions to the Fruit Tree Centers program and Commercial Sales. Stoeckley said he reduced selling costs by about 6.5 percent and nearly doubled company revenues during his five-year tenure, two main areas of concern for the lender.

John Stoeckley

He was, however, often at odds with Clay over the way things ought to be run. They butted heads over compensation structure for the Sales team, and a working environment Stoeckley felt was less than supportive of innovation and creativity. In 1990, John said he was let go out of a "need to cut costs."

Said Clay: "My impression was John's artistic talent exceeded his managerial talents. The success of initiatives he instituted at Stark were a mixed bag — some were fruitful and others were not successful. After five years, I concluded his performance as head of sales and marketing wasn't a good fit. I still count John as a friend and think he considers me a friend as well."

At the time, the Garden Center made a relatively small contribution to annual revenues: $300,000 to $400,000, or two to three percent of gross sales. It's a lot more now, and it is a profit center in its own right.

"Still, we were pleasantly surprised by the volume," said Clay. In tandem with Fulfillment, the now-diversified company had a better hedge again downturns.

Ginny Andrews, who was instrumental in the Garden Center's success, said the store draws visitors from all over the U. S., many of them devoted catalog customers. Even after retiring in 2009 after 48 years, she comes in three days a week in the busy spring months to help run things. She is nothing less than a smiling fireball of energy and positive attitude.

Said a former co-worker, "Perhaps the Production Department should figure out a way to propagate Ginny."

—1980-1991—

In the same year as the Garden Center opening, Luther Burbank was posthumously inducted into the National Inventors Hall of Fame®. Clay Stark Logan was asked to accept the award, with Jim Law, Vice President of Production. A bigger honor would be hard to bestow.

A 1983 ad for the new Garden Center

People came and went; fewer Starks were there than ever. After ten years or so as the last research horticulturist in the family, Paul Stark III lost interest in the business. Faced with an irreconcilable marriage, he left Missouri and moved to California, where his mother was. He died in 2004 of a brain aneurysm. Sadly, his grown daughter died shortly thereafter.

PIII's dad, Paul Jr., while retired, came and went, too. "When I was running the company, Paul Jr. would often wander into the office, possibly under the influence," recalled Clay. "He would rummage through files and pick up books, papers, etc., even though he was no longer employed there."

In 1987, Terry Stark (no relation) celebrated her 10th year at Stark Bro's, where she started in the Commercial Division. A self-described "tomboy right off the farm," she holds very warm feelings for the nursery. "When I came here I was very shy. I now talk to clubs and give tours. The nursery has allowed me to grow as an individual, and I have gained a wealth of knowledge, for which I give Elmer [Kidd] full credit," she said, tears of gratitude rolling down her face.

Then in an instant, a smile appeared: "The one thing I have learned: before you work on a tree, look for hornets."

Back: Clay (left) and Walter; John (seated) and a rare photo of Paul Stark III (kneeling)

Terry was Stark Bro's first female grafter, and if there were ever a personification of the spirit in this nursery, it is this extraordinarily dedicated woman. There are so many others like her; these are very special people.

The Commercial Division, where many employees like Terry start out, offers a complex catalog of scion-rootstock combinations to meet orchardists' needs. But despite its technical savvy, Stark Bro's has historically been ineffective in a crucial market: California. Its import regulations are tough, so there was not a big push to ship there, but Clay wanted to change that as well. In 1988, he bought a 245-acre California operation named Agri Sun Nursery.

Through Agri Sun, Stark Bro's began supplying trees for high-density fruit production. West-coast fruit growers were cramming more and more trees onto their property in the hope of turning a higher profit. Clay warned that planting trees too closely can create a jungle, and "nobody can grow fruit in a jungle."

By finding new markets and squeezing more profit from existing systems and personnel, Stark Bro's emerged from years of crisis. A better economy boosted Clay's two main strategies for capturing new retail mail-order business: diversifying the customer base and broadening the product mix.

By 1991 — incidentally the company's 175th anniversary — things had finally settled down. Annual gross sales are around $15 million, with mail-order catalogs and 800 phone numbers bringing in more than 400,000 orders a year. The state of affairs was certainly better in 1991 than it was five years ago, but it was not anywhere near as flush as it was 15 years ago.

One more event colored the end of the decade: Paul Jr. got married to Wife #4, his last, Lois.

The Alar Scare

Few apple growers will forget the February, 1989 "60 Minutes" segment about a growth-regulating chemical called Alar, featuring an apple marked with a skull and crossbones. The report stemmed from a Natural Resources Defense Council study that concluded Alar posed a cancer threat, particularly to children.

Public outcry was immediate; actress Meryl Streep took up the cause, demanding Alar be banned. Some still question the science behind the report, but today, Alar is applied only to ornamental plants. Clay Logan, who wrote a scathing letter to the editor of the St. Louis Post-Dispatch, slammed the media for elevating the scare out of proportion. The kerfuffle came and went, and in the long run, was just a blip on the apple industry's radar.

Stark Catalogs Through The Years

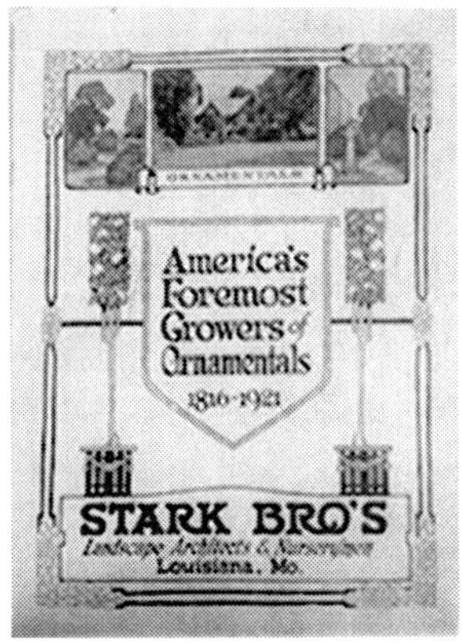

17

1991: The 175th Anniversary

Changing Customer Tastes

The Wonderful World of Elmer Kidd

17 | THE FRUIT OF 175 YEARS

1991-1994

The 175th Anniversary

Little did James Stark realize when he filled his saddlebags with apple scions from his father's Kentucky orchard that he would create such a legacy.

Clay (left) and Robert Dolibois, Executive Vice-President of the American Association of Nurserymen

The 175th anniversary celebration was held on September 7, 1991. The *Louisiana Press Journal* marked the occasion with a 28-page special supplement.

For this special "birthday," famous *TODAY* Show weatherman Willard Scott held up a Red Delicious apple on national television and said, "I had the best breakfast in the world, cereal with raspberries on it. I got 'em from Stark Bro's and they're the best raspberries in the world. Call 'em up!"

The segment caused a flood of phone orders, and it took a while before Barb Jenkins and her group figured out which variety he meant. Customer service records show that Mr. Scott did personally order from Stark Bro's.

The 175th anniversary cake. Left to right: Michael "Blake" Logan, Bethany and Benjamin (Walter's children) and John (Clay's son).

The old Stark Cabin was decorated with family antiques, and there was the requisite cake. Speakers recounted the story of James Hart Stark and how the Delicious apples came about.

A tree was planted (above right) to mark the event, along with a time capsule. No one knows what happened to it; some dismissed the idea that there ever was one. The tree is gone and so is the marker that may (or may not) have identified the spot.

Changing Customer Tastes

Some of the discussion on that special day was about how Stark Bro's had survived for so long. One of the answers came from Research. "Unique sells," said Joe Preczewski, Director of Research at the time. "I can come across the best-tasting strawberry variety ever, and if there's nothing unique about it, other than it's just the best-tasting strawberry. It's not hard to sell commercially, but it's hard to sell to the homeowner.

One of Stark Bro's huge greenhouses

"Now, give me a strawberry that has red and white stripes - and it could taste very inferior to the average strawberry — and you would sell a lot of those."

Note that this is completely the opposite of the reaction to the Red and Golden Delicious apples when they were introduced; they were rejected because they were too different! Times do change.

Joe also worked with many independent breeders. The company acquired exclusive mail-order rights to Floyd Zaiger's two new plum-apricot hybrids, trademarked as Aprium® and Pluot®, completely new fruits that are relatively commonplace in grocery stores today.

'Colonnade' was another novelty. It was introduced in 1994 and seemed like a great idea. "The tree grows like a cornstalk," said Clay Logan. "There's one central shoot with little fruiting spurs right on the trunk and leaves every half-inch or so. In essence, you can build a flowering wall, with trees one- to two-feet apart."

Colonnade was a clever space-saving concept, but unfortunately didn't have the flavor "wow" to propel it to stardom. Stark Bro's is revisiting the variety because of the renewed interested in small-space gardening and urban farming.

Back then, Research came up with a few new and different packaging choices. Retailers could buy the traditional bare-root stock to sell either "as is, or get it already potted. Root-wrapped plants were also available in plastic bags.

The Wonderful World of Elmer Kidd

In 1965, Elmer Kidd earned his first paycheck hauling drinking water to Stark's field hands. Since 1991, he has supervised its orchards and greenhouses as Chief of Production, and is a human encyclopedia of horticultural knowledge. His son, Billy, works with him.

Elmer is hard to categorize. He is brilliant; he is homespun; he is a man of deep faith; he is hilariously funny. There are stories about his stories, and he has many. He also has an opinion about most everything.

The ingenious bottomless 4x10 pots

On the Nut Program: "As a young man I had a dream (a co-worker called it a 'windblown idea') to bring hard-to-propagate trees like nuts, pawpaws and persimmons to the home grower. I built the program together with Paul Stark Jr. and Lou Brown, Cameron's dad. Putting the trees in a 4"x10" pot with holes in the bottom "air-prunes" the trees and forces the feeder roots to grow within the pot, giving them a much better chance than a tree that was dug up from the field. It was hard to convince the higher-ups, but I did, and now there's the Stark EZ Start® Program, which has grown exponentially."

On the future of Stark Bro's: "Stark Bro's will be here in 200 years if it holds to its practices, yes. You can buy a dead tree at box stores, sure. But we have a broader offering, and the relationship continues after the sale."

On the bankruptcy: "It was like getting a fence post pushed right through my heart. That was probably the most agonizing time in my life."

On his capability: "When Cameron first bought Stark, I oversaw all of it for 8 or 10 years, the warehouses, the fields, the shop, everything but sales. It wasn't an easy thing to do."

On fruit licensing: "Most of the new varieties are coming out of the universities, and they license growers to grow them. They don't just want to give the variety to one nursery; they want everybody riding the bicycle for the royalty money."

On childhood obesity: "Grocery-store apples are chosen for color and shelf life, not flavor. That's why kids would rather eat a Snickers® bar."

On his favorite apples: "Best variety in the last 10 years at Stark is Gala. Other good apples are Jonagold and Fuji. Arkansas Black: apple salad. Jonathan: caramel apple. Jonathan and Northern Spy: pie apple."

On persistence: "Poplars and willows, anyone can do. PawPaws, no. What John Q. Public doesn't tell you is that when you buy from a box store, it ends there. Here, we follow the ball over the fence, all the way to the end."

On work: "I've gotten to do what I love, my whole life. As long as the owners will have me, I'll be here. I didn't have to come to work this morning; I got to."

Governor and Mrs. Nixon of Missouri, planting a tree with Elmer Kidd.

During another visit on October 19, 2012, Gov. Jay Nixon presented Elmer with the Governor's Award for Agricultural Achievement in recognition of his work on the 4x10 pots.

On technology: "Back then, the computer took up about the whole computer room and cost THIS MUCH. Clay wanted an immediate inventory system on computer; that was a dream 30 years ago, there was just no way to scratch the itch. I'm not a computer person; if we're talking what do you like better, Star Trek or Gunsmoke, I'm definitely Gunsmoke. Laptops, iPads and wifi have since fulfilled Clay's dream."

On the younger generation: "They come along, you know, the longer they're here they start to absorb the values. There's a saying here: 'Once you get moss down your shirt, you're always going to be here.'"

On when to plant: "I'd recommend spring, but would not try to persuade you not to plant in the fall. The reason is that if we have a harder winter than normal and we plant in the fall, the winter pulls the soup out [the plants]."

On "the budders:" (The summer the nursery closed and before its assets were sold, 1 million rootstocks needed budding, and a crop of 800,000 fruit trees in the field needed tending. Elmer picked hourly workers to tend the crop, and five volunteers worked without pay for two weeks.) "Everyone had a winning spirit, working shoulder to shoulder. I remember it like no other time."

On naming apple varieties: "I named a couple of apple varieties. One was called 'Sweet Pea,' the other 'Little Jewel,' which was a really good apple. Both flopped, which convinced me I should stick to growing."

On customer samples: Stark used to be about the only place where you could go with your 'creations.' Now people send fruit just for us to identify. 'Is this a good one?' Depends. One person can look at something and call it a burr. Another person calls it Velcro®."

The Importance Of Sampling

Familiarity with the Stark name often prompts amateurs to send fruit to the company, hoping to peddle "the next Red Delicious." Most of the samples are unremarkable or aren't anything new, but there are dozens of legitimate samples worth investigating. Evaluating samples is a lot of work, and it's expensive. Chance finds require extensive testing, virus-indexing, patenting and marketing.

"You have to kiss a lot of toads before you find a handsome prince," said Clay Stark Logan, the last family member to serve as president. "But finding one is what sets Stark apart. When you look through other catalogs, it's pretty much the same old thing: double red, red Rome, red Jonathan. Only two or three other wholesalers offer anything in the way of patented, different fruit varieties."

There is an atmosphere of a never-ending treasure hunt each time a sample box is opened at Stark Bro's. Every sample, no matter how unpromising it may look, is tasted and examined, because inside that box could be the next Golden Delicious. The owners also remember that when the original Delicious turned up, the Starks believed they had at last found the perfect, unimprovable apple. Yet twice since, freak genetic mutations have given them even better varieties, like Starking® Delicious and the semi-dwarf Starkrimson® Delicious, both bud-sports (chance mutations) of the original Delicious. If these varieties aren't noticed and preserved, they can be lost forever. One can only ponder what might have already been lost.

18

The Sale

The Bankruptcy and The Sale, Again

The "Budders"

After the Auction

The Next 200 Years

18 | THE END GAME, AND ONWARD

1994 - 2016

1994 was the 100th anniversary of Red Delicious. Since then, Red Delicious production has declined by about 40 percent, but it is still the most common apple in the U. S. In 2011, growers produced 63 percent more Red Delicious than its closest competitor, Gala, and HoneyCrisp is coming up fast.

In 2015, the biggest export markets for the Red Delicious are in Southeast Asia. China, which now produces more apples than any other country in the world, has been a major buyer of Red Delicious from Washington orchards.

What used to be is not always what will be.

The Sale

Clay, Walter and their dad, John, had explored selling the company several times in the '70s and '80s — once to Burpee, the famous seed company.

"We danced with them, met their folks," said Walter. "Dad just never liked the idea, and there was no formal offer. We were approached by another wheeler-dealer in the hort field who had a lot of contacts, but we just didn't feel like that was a good long-term deal, either."

The family formulated some conditions for a possible sale: if it was good for the stockholders and gave the employees a future, they would go for it.

One of the lesser reasons for selling was the wake-up call of the 1993 flood. The Illinois levee had held this time, but what about next time? If it broke, the company would be finished. Clay and Walter were troubled by that possibility.

Sometimes the answer is in your own back yard. Stark Fulfillment Services, headed up by Jack Alexander, had started out doing contract work for Foster & Gallagher (hereafter known as F&G). They were an employee-owned company, selling garden and gift products via catalog. Their circulation in 1999 was more than 300 million, including catalogs and other direct mail pieces, along with a house file of more than 5.3 million names. Gross sales: a healthy $340 million.

F&G's titles were familiar ones: Spring Hill, Michigan Bulb, Gurney's and others. It was Stark's biggest fulfillment customer, and it was also the company to which Clay and Walter would ultimately sell the family business.

From F&G's point of view, it was a perfect deal. Owning Stark Bro's would be far cheaper than keeping up the fulfillment contract. The established business had the buildings, the cold storage, the people, everything — all in one place. F&G could put their money in, and take out whatever they wanted, almost like a vending machine. As part of the sale agreement, Clay and Walter were each offered a five-year contract to manage sales/production/operations and Mail Order, respectively. Jack Alexander remained with Fulfillment.

"The opportunity they offered our Fulfillment division was tremendous, and I think a lot of the employees would have benefitted," said Clay. "F&G was getting bigger, and the employees could have gotten bigger with them. At the time, it looked like a good deal for everyone involved."

Familiar Foster & Gallagher catalogs

Some of the employees were in the camp that thought it was F&G's idea to buy — but not Clay's idea to sell — and that an offer was put in front of him only after F&G worked with Jack Alexander to influence Clay. The two were lifelong friends, and one could very likely sway the other. Clay has said that no such discussions took place.

Jack Alexander was viewed as the "darling" of Stark by F&G. He spent a good part of his time trying to build rapport with them, portraying himself as "your next Stark Bro's president," even though he was well aware of Clay and Walter's five-year extension contracts. Internally, many thought Jack shouldn't be engaging in a power struggle with his long-time friend.

There are three stories about the sale announcement. Story One goes that Clay called everyone into the offices, and stoically told the group that the company had been sold. He laid out the facts, talked about the inevitability of change and carefully explained his rationale for selling. Clay doesn't remember.

Story Two says Walter began the announcement, but broke down and Clay took over. Yet another version says John Logan, then Chairman of the Board, came in to make the announcement. The most peculiar aspect of this is that neither Walter nor Clay remembers for sure who made the announcement, a real testament to the weight that must have been on them both that day.

Whatever the story, it was a surprise to some; others wondered what had taken so long. To everyone, it was hard to comprehend how a company run by six generations of one family over 178 years would be no more.

Walter is philosophical about the decision now.

"It was wise to sell the company given the climate, and where we were," he said. "There were weeks and months when there was contemplation of filing for bankruptcy. It would be better for the stockholders and employees to sell. When you walk away from a situation like that, it's the people and the relationships that stay with you. Without [them], we never would have survived as long as we did."

The employees had a Profit-Sharing Retirement Trust to provide a supplement to their Social Security benefits. Thankfully, those monies were distributed to all eligible employees when the company was sold.

Many people think the Logans got all the money from the sale, which wasn't so. They were the majority owners (51 percent), but there were a number of other shareholders (Stark cousins); according to Clay. "One or two took issue with my and Walter's agreements to stay on, but in the end, everyone voted to sell."

Clay added that most of the Stark Bro's stock was held by various trusts that benefitted about 20 different family members.

"I've never had second thoughts about [selling the company]," he said. "Besides giving the sale proceeds to the stockholders, there were a number of strong, objective reasons we decided to sell. Probably the biggest reason was the significant diversity of ownership, especially among family members who had never been directly involved the nursery."

In the conference room, the walls were once lined with a procession of portraits of all of the deceased "Stark men." F&G removed them; no one knows what happened to them. The precious silver Wilder medals also disappeared. (A footnote: the medals have been replaced and hang in the lobby of the corporate offices.)

And so life began under the ownership of F&G. In 1995, three acres' worth of production greenhouses were constructed at Stark Bro's to grow products for F&G's sister companies and fulfillment clients.

At the height of F&G's ownership in 1996, Stark Bro's employed more than 700 people, but budget cuts were starting to rear up and there was a series of painful layoffs. Departments were consolidated and some duties were relegated back to F&G's home office in Peoria, Illinois. Sales had dropped, but only after the F&G brass had made huge catalog changes Walter saw as "detrimental." He pushed for more testing, and the "old" catalog performed just fine, so they reverted back to the old model.

It didn't matter much, because F&G was focused on fulfillment, not the nursery, and the commercial area also suffered, sales-wise. About 1.25 million trees were sold in 1999. More cuts and layoffs ensued, making many of the Stark employees jittery.

Despite poor Stark Bro's catalog sales, F&G was growing faster than a beanstalk — thanks to the Michigan Bulb sweepstakes-marketing scheme. The offer: buy any horticultural products at entry-level pricing (e.g., 100 tulip bulbs for $9.95) and be entered in a sweepstakes. The program generated hundreds of thousands of new customers for the F&G database every year. Moreover, these customers were an excellent mailing list for other F&G catalogs, and for renters of the Michigan Bulb names.

But then the Federal Trade Commission started cracking down on sweepstakes-by-mail. F&G stubbornly refused to test alternative marketing methods, and kept riding their gravy train while the FTC formulated its tight regulations, causing Michigan Bulb to wilt. Negative publicity pushed it further underwater.

Very near the end of his employment contract, Clay tired of being a puppet on a string for the "corporate muckety-mucks" and was ready to pull the plug. Budget-cut rumors were also floating around. In February of 1999, he left, and so did Walter. Jack Alexander was promoted to VP and General Manager, reporting to F&G.

The following year, Congress passed the Deceptive Mail Prevention and Enforcement Act, which took effect on April 12, 2000. It would signal the end of F&G.

The Bankruptcy and the Sale, Again

F&G shipped 17 million packages in the year 2000, generating a staggering amount of income. But after the sweepstakes crackdown, Michigan Bulb began hemorrhaging money as its response rates plummeted. F&G stubbornly hung on, an enormous mistake because well over half of their revenue and profit came from their horticultural titles. Michigan Bulb's collapse had a domino effect on the other catalogs. Implosion was imminent.

F&G had two choices: either seek a new winning formula for Michigan Bulb, or dump it and cut everything else back. There were no investors to save the day; remember, the company was employee-owned. The Board failed to lean on management to make the hard decisions, and predictably, the company failed. More than 3,000 jobs were lost.

The timeline looked like this:

June 13: Robert A. Ostertag Jr., CEO of F&G, apparently saw the writing on the wall, and resigned.

June 22: Another of F&G's companies began closing some of its departments, and people's responsibilities were shifted around at Stark Bro's. Rumors started ramping up. The head of F&G's HR department called Marian Kuntz, head of Stark's HR department, with a heads-up.

June 29: F&G shut its doors in the Midwest, Canada and the Netherlands, saying it could not continue normal operations while restructuring was in progress. Stark Bro's was included. At 8:00 am, the employees were sent an urgent e-mail to meet in the lunchroom in five minutes. Jack Alexander walked in.

"We're shutting down," he said, but could not finish. He sat down. Marian Kuntz took over and handed out a prepared document to the stunned room. The letter spelled out local unemployment resources and how to file, as well as a personal note from her to the employees:

> *"I am having a difficult time coming up with the right words, probably because there are none. However, please know that I have enjoyed working with you all and appreciate you allowing me to be a part of a great group of people — a second family."*

A seismograph could have picked up the shock waves. Clay and Walter said later that they never saw it coming.

"The hardest day of my life was June 29, 2001, when I had to look the Stark employees in the eye and tell them the company was shutting its doors," said Alexander. He said he cried when he went out to tell the field workers. Later that day, he called Clay and together they took two truckloads of pictures and documents to the Louisiana Area Historical Museum so they wouldn't be lost, as the family portraits and Wilder medals had.

Many of those who were fired had worked there for 40 years or more. There was no warning, and no recourse. It had yet to sink in that there would be no severance pay, and that their pensions, 401ks, vacation time and health insurance — seven year's worth of accumulated savings and benefits — were gone.

Beyond that was the apparent end of an American institution. All the history and landmark discoveries that the Stark name represented would be dust in the wind if the company dissolved. Worse yet, the employees knew they had done nothing to deserve this fate. Workers describe that day as one of the darkest of their lives.

Ricky Blackstun

"Grown men were crying," said Ricky Blackstun, one of the long-time employees. "It's the only thing many of us know how to do, and we were rattled."

"It was doomsday in Louisiana, Missouri," recalled Walt Gilbert, publisher of the *Louisiana Press-Journal.* At the Stark Bro's corporate offices, the flag was flown at half-staff.

Ginny Andrews was taking that day off, preparing for her son's wedding, when she got the word: Stark Bro's Nurseries & Orchards Co. was closing its doors. Andrews, a 40-year veteran of the company, was among the casualties.

"I just crumbled after I got home from my son's wedding. I stayed home for four months, walking the floors, wondering what was going to happen," she said.

The bank required that the company remain open to protect the crop for a future buyer, and so Elmer Kidd was asked to pick out 10 people he thought could keep the crop going. About 73 others were retained to run the rest of the nursery and fulfillment operations. The remaining employees, about 470, were permanently laid off. Within an hour, the building was almost empty. People stood in the parking lot, just staring at the hills.

It was hard not to be bitter; the employees had worked hard for F&G and were left with nothing. The future had looked bright, there were lots of changes, lots of money, but people became numbers and lost their identity.

Marian Kuntz, Stark Bro's HR and Finance Manager, was retained to wear a number of different hats. She tried to keep the "old-timers" in the loop, saying, "hang in there, we're going to be bought, just hang in there." Marian was their hopeline, helping them with unemployment, providing them with records and documentation and giving them desperately needed moral support.

July 2: F&G announced they were filing for Chapter 11 bankruptcy protection. No bank would finance them, and so they opted to sell almost all of their assets, including Henry Fields, Gurney's, Stark Bro's and 12 other entities. The only catalog to escape liquidation was Walter Drake, seller of talking mounted fish and handy kitchen gadgets. F&G had literally been yanked out by its roots. According to the court petition, F&G listed more than $100 million in assets, and more than $100 million in debt.

Said one orchardist on a garden forum page: "From the oldest nursery in North America to Chapter 11 in three short years! Good Lord, this was the nursery that introduced the 19th century world to the Red Delicious apple!"

A frenzy erupted when it was announced that anyone who had placed orders from F&G-owned companies (and all other customers who were expecting replacement orders the following season) would be left hanging. That included thousands of orchard growers who were loyal to Stark Bro's, a rock-steady business that had been a reliable wholesale supplier for decades. A long-time switchboard operator, Vera Murr, deftly fielded a deluge of phone calls from confused and concerned Stark customers.

There was one other problem: there were still 1 million rootstocks in the field to be budded, and 800,000 trees to be cared for. Trees usually are budded from June to September, and by August 22, the work still hadn't been done, pending direction from the bankruptcy court. Between June and August, the bank was funding the Stark fulfillment business, but not the nursery. No one did maintenance on the coolers, power lines, the roof, nothing. There was no money for anything, no money to pay anyone to save the trees.

The "Budders"

The layoffs left some questioning their sanity after a few weeks; they were workers, and they needed to *work*. Moreover, they missed the nursery, the feel of the earth, the history, the blooming spring, the harvest. They waited, and prayed.

Terry Stark was one of those with cabin fever. "We're not doing anything," she said, "so let's just go pull weeds in the field." A group of five — Terry, Barb Brewer, Bill Griffith, Lois Schneidler and Ella Patrick — got permission from Jack Alexander to use the company van and go over to the Illinois side for a few weeks. Workers over there were getting paid; the volunteer group was not, and did not care. They knew their only chance at getting their jobs back hinged on growing those trees.

Richard Owens had a nursery in Bloomington, Illinois, and was interested in buying Stark Bro's He offered to pay some workers to get the budding done in case he won the auction bid. So from that point on, they all got paid, and Elmer Kidd was instrumental in getting the job done.

"In late August, we had a meeting, about 100 of us, out at the end of the row, and I told the crew, 'I don't know how we're are going to do this, but with God's help we are going to get this done,'" Elmer recalls. They worked quietly, with little instruction — six days a week, 10 hours a day — until it was done. The weather held, and budding was finished October 3, an impossibly late date.

"I've endured a lot of things in my life, but I can't remember anything like that. It's something I'll tell my grandkids about," Elmer said

The story was eventually told by the late Paul Harvey on his famous "The Rest of the Story" radio segment. It is those volunteers who so perfectly illustrate the character and commitment of the "other" Stark family.

July 9: Ginny was elated when Jack Alexander, who had become VP and General Manager, asked her to come in and close down the Garden Center. "Me and my assistant made a pact that we would sell every single thing. We had six weeks. We wheeled and dealed until there was nothing left, said Ginny," tearful at the memory. The Garden Center closed its doors on July 28.

July 26: F&G hires a New York PR firm, and Lawrence Adelman as interim CEO. His salary is $21,000 a *week*, infuriating Stark employees who had been told, "we are out of money."

An outside company bought the rights to ship fall bulb orders for F&G out of the Spring Hill Nurseries plant in Ohio. That meant people would be needed to take calls and process orders, which meant Marian Kuntz would get to stay beyond her July 28 contract-end. In addition to 47 relieved Stark Bro's order-takers, there was an OK to hire 85 more budders to try and save the crop.

August 2: An auction date of August 21 was set for F&G's horticultural titles.

Katherine (Cameron Brown's mother), who had rented growing land to the Starks for years, was now renting it to F&G. She received a letter from the Delaware bankruptcy court that F&G had gone under. The assets were to be sold, and that she might get something out of it, but she didn't: she lost $25,000 in rental fees. Katherine called Cameron about it, and that is how he found out that Stark Bro's was for sale. Cameron casually proposed the idea of buying it to his business partner, Tim Abair, and they started doing some due diligence.

The men traveled to Louisiana to interview key staff and spent weeks sifting through contracts, financials and historical documents. It was turning out not to be such a lark, after all. They decided to attend the auction, which had been delayed until Tuesday morning, September 11.

August 27: Jack Alexander acknowledged that he convinced the bankruptcy managers to sell him the fulfillment business. That was not in Stark Bro's best interest, but it was certainly in Jack's. "It was just a good business deal," he said.

The bankruptcy court approved the sale of Stark Bro's Fulfillment to Jack's Lexton Group before the auction. Jack bought it directly from F&G and went off on his own. He kept the Stark Bro's Fulfillment name because he felt the credibility would give a leg up to his new business.

About 40 employees went with him to work at a large building on Georgia Street — a straight one-mile shot down the road from Stark Bro's Nurseries & Orchards Co.

An "old-timer" who requested anonymity, said, "I didn't know Jack had bought Fulfillment. Didn't bother me, but it bothered management. People said Jack slipped in the back and pushed everybody else out. Jack and Clay were frat brothers, but Jack never told him what he was up to. One fella said, 'Jack sold his buddy out.' I would have to agree with that." Clay has said that there are no hard feelings.

September 11: Tim and Cameron drove past O'Hare Airport on the way into downtown Chicago. Pilot that he is, Cameron was always looking up to see what kind of jets were taking off. Strangely, that morning he didn't see a single plane.

They continued on into the city, slowing down for a light at the busy intersection of Jackson and LaSalle. Normally, this would be rush hour, but there were only a few policemen and a handful of people there. It was unsettling at best, so Cameron turned on the radio and they heard about the terrorist attack at the World Trade Center.

Planes grounded at O'Hare Int'l. Airport on 9/11/2001

Tim called the firm holding the auction and asked if it was still going to proceed. Shockingly, they said it was, because so many people had come from all over the country to participate. The site was changed, however, from a building that was about 30-stories high to a 7-story building.

Stark Bro's Nurseries & Orchards Co. was Auction Lot I. There were about 50 people there, an industry Who's-Who, with their own legal teams. The auction went on into the night. Jim Adashek, the attorney for Tim and Cameron, thought it had been poorly run. They didn't know if they had won or not.

It turns out that they tentatively won the bid, but intense negotiations went on for weeks afterward. It was never a sure thing. Then a nice monkey wrench: After winning the bid, later that day the pair learned that there was to be an enterprise bid for all of the assets. (If someone bids more than the sum of the individual lot bids, they take the entire enterprise). So Stark Bro's was still hanging by a thread.

An attorney approached Cameron and much to their dismay, told Tim and Cameron that his client, Gardens Alive!, had enough money to win the enterprise bid. Their hearts sunk, but the choppy boat ride was not yet over.

That same attorney then offered to sell Stark Bro's back to Tim and Cameron for what they successfully bid for it, because the folks at Gardens Alive! wanted Stark to continue to raise trees for them. They all shook on it.

Gardens Alive! indeed won the enterprise bid, paying $10.75 million for Stark Bro's. Now all they had to do was keep their word.

The parties jockeyed back and forth over the condition of the inventory in the field, the price, the down payment, customer data and the technology. It was like a waking nightmare for the Stark workers, because they weren't getting paid and didn't know if they would ever get their jobs back.

Bruce Griffith at Mercantile Bank (in Louisiana) was the only lender who would give Tim and Cameron the money to buy the company. As of October 10th, it was starting to look better, but there were still anxious moments.

At about the same time, the money from Richard Owens was running low. The recall-hiring was finished, and so Marian Kuntz was finally laid off. She stopped into the office a couple of times a week, checking for any news.

Jack hoped he would be named president of Cameron and Tim's new company. He had been angling for it for a while now, and had given Cameron a proposal. A group of employees (led by Kim Young, then the Greengoods buyer), who viewed Jack Alexander as a Machiavelli of sorts, met with the new owners in an attempt to convince them that they could do the job themselves. Cameron and Tim sent Kim the proposal Jack had drawn up for himself.

In the meantime, Cameron, Tim and Jack had a closed-door meeting that set the company's email server afire. The employees could hardly stand the suspense. The meeting was to tell Jack that he would not be named president.

Kim Young had indeed convinced the owners that she could do the job cheaper and better than Jack. She was eventually promoted to VP and General Manager. Tim offered Marian the Human Resources Manager position and asked if she would do a little accounting, an offer she gladly accepted.

When the money indisputably ran out in early October, Jack paid the wages of the budding crew to finish that job. That gesture allowed Tim and Cameron to secure Worker's Compensation insurance and get set up to pay employees. In the eyes of some, this redeemed Jack a little. Then in December, Jack billed the new owners for the cost of the "gesture."

The deal was finally closed on October 26, 2001 in Covington, Kentucky. Gardens Alive! had kept their word. The wire transfer went through, and the rest, as they say, is history.

Stark's Chief of Production couldn't have asked for more. "It was a fairy-tale ending," Elmer said.

Jack went on to make Stark Bro's Fulfillment Services a very successful company. He retired in 2009. In 2012, the company name was changed to Fifth Gear, which was recently sold.

After the Auction

Cameron and Tim called everyone down to the shed and introduced themselves as the new owners. The employees were ecstatic, but there was no time for a celebration just yet. There was too much work to be done.

First off, as CFO, Tim wanted the Garden Center financial records, and asked Ginny Andrews for them. She had to tell Tim that Jack Alexander had taken them all. She smiled, Tim smiled and they agreed they would find a way to reconstruct the records. (Tim was heavily involved at the beginning, re-establishing a finance and accounting system. Now his role is one more of oversight.)

To their unhappy surprise, Tim and Cameron discovered a small strip of land through the middle of the property that wasn't part of the auction bid, so they had to buy that back at a premium. Then they found out that Jack Alexander had taken the software for the order system, including manifesting, shipping and handling, along with most of the company records — not just the Garden Center's.

There were thousands of trees to ship and no way to ship them. They needed the old system for a while until they could get one of their own, and so they asked Jack. In true Snidely Whiplash fashion, Jack charged them a premium for the next year to use the old system. A double-whammy was not a good way to start.

A skinnied-down Stark Bro's office staff, hand-picked by Kim Young and Elmer Kidd, came back to sad, empty desks and file cabinets, no computers and no phones. It had all been lost to the bankruptcy, and pandemonium resulted.

The department managers took the reins, knowing they had just six weeks to produce a spring catalog. It was grueling, but a full-court press was necessary because then, 98 percent of the business came from the catalog. Pricing, layout and printing in such a short period "was the toughest, but most rewarding thing, I've ever done," said marketing specialist Lita Eatock, who had started in 1978.

The first catalog under the new ownership mailed on December 28 to about 2 million house customers. There was no fall shipping in 2001 for Stark Bro's, but they still fulfilled orders for Gardens Alive! and all of its companies.

The employees came back to work in pieces: the budding crew, then more workers for harvest, managers for the office, inventory and warehouse. Jack's new company, Stark Fulfillment, processed orders through their computer system, and sent over the pack-slips to print and ship.

F&G had burned vendor bridges because of their bankruptcy, and Stark Bro's had to prove themselves all over again, to both commercial and retail customers. Many had been stiffed the year before when F&G ceased operations without shipping their orders. Stark mended that fence by shipping out all 5,600 missing F&G orders, free of charge.

When Stark Bro's catalogs for 2002 started appearing in mailboxes over the winter, people wondered whether to give the company another try. Factor in the post-9/11 anthrax scare, and all the employees could hope for is that people would open their mailboxes. The first year was most definitely the toughest.

In 2002, Cameron's son Kyle joined the company as Chief Operations Officer. At the time, the priority was to keep the doors open, not to charge ahead with big updates. But within a year, Kyle shined a 500-watt light on the need for technology improvements.

His number one goal was to apply technology to a company that was 100 percent paper. It was also obvious that they would need a shipping system to accommodate the new e-commerce business, and that the old brand was "tired" and needed to be refreshed as well.

Something had to be done about the order-processing system; they couldn't go on renting it from Jack Alexander forever. Kyle made the decision to develop an in-house software platform for tracking orders and inventory, along with the basics the company needed to operate. The technology he developed was called URSA, which is still in use. (Ursa Major, for which the system was named, is the "bear" constellation, a perfect sync for the Stark bear logo.) That initial project was outsourced to a St. Louis IT consulting firm. In about seven months, they came up with an initial bare-bones system.

Kyle was the driving force behind transforming the company into a modern, technology-driven enterprise, though he had to drag it, kicking and screaming, into the 21st century. "We had invested an awful lot of money in computers and wanted it to pay off," said co-owner Tim Abair.

The company was now on par with their competitors from a systems standpoint. The website still had a marathon to run.

The Stark Bro's Website

This one-font version was the first time the world saw StarkBros.com, in 1999 (right). Visitors could search for products or request a catalog, but there was no shopping cart.

In April of 1999, the Stark online "store" was outsourced to mySeasons.com (now defunct), and then largely abandoned. On April 1, 2002, the only thing on the website was this:

"Thank you for your interest in ordering from Stark Bro's We are finishing the work on our new website, and will be able to process your order on-line in a few days."
On June 1, it read the same.

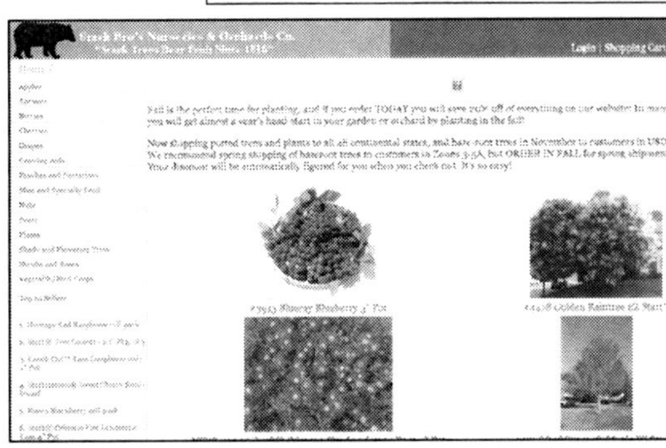

By September, there had been an overhaul, but no e-commerce until January of 2003, when just two products were offered. (Stark Bro's was very late to the e-Commerce party.) By 2004 (right) there was something beginning to resemble a website, thanks to Kyle Brown's efforts.

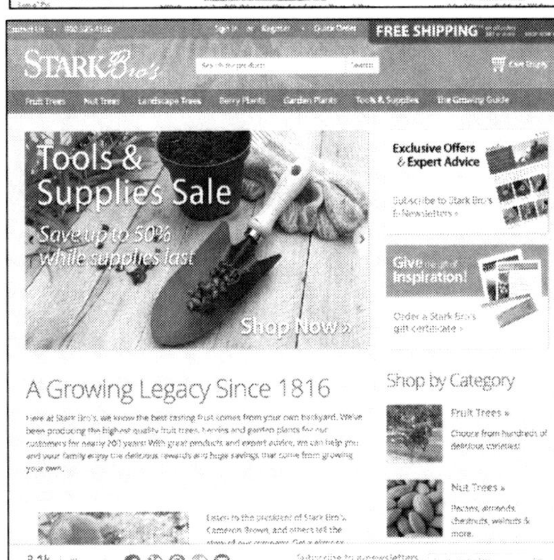

In 2006, there was a huge leap forward in the website design, and by 2009, Jacob DuBray had developed the design into its current state (right).

Today, Stark Bro's website is ahead of most of the nursery industry in their online capabilities and has one of the most comprehensive gardening resources on the Web.

Busy employees were handling nearly 4,000 orders daily, about half for their own business, and half for Gardens Alive! and their familiar catalogs (Henry Fields, Gurney's, and others) that used to be owned by the now-defunct F&G. Cameron gave full credit for the company's rapid progress to its "loyal, experienced workers." A little more than a year after Tim and Cameron bought the nursery's assets, two-thirds of the lost jobs were restored.

By the spring of 2003, the company was definitely getting on track. Customer service was back in the hands of Stark Bro's, under the management of Debbie Epley, another Stark "old-timer."

"At this point we're rolling. The first two years were really a lot of work," said VP and General Manager Kim Young. A capital investment of close to $3 million from the owners did not hurt matters.

The undaunted IT department started developing URSA with a vengeance, with Bobby Birks assigned to work on the system full-time. Kyle's goals were now focused on improving the website. An old college friend of his designed the initial site; Kyle himself would make small changes to it, like new offers and promotions on the Home page. It wouldn't be until several years later that Stark Bro's got a "real website."

As the IT arm of the company grew, staffing became a challenge. Being in a rural area, there was/is a labor shortage of IT specialists, so Stark Bro's opened a satellite office in Lake St. Louis in 2008. In searching that talent pool, they found Jacob DuBray, who was largely responsible for developing the company website after Kyle was gone.

Kyle David Brown

Kyle was promoted to Chief Marketing Officer in 2006. By April of that year, there was a huge leap forward in the website design and function, which remained fairly static until February of 2009, when the current green color scheme made its first appearance, courtesy of Jacob DuBray. The website has gone from generating 0.2 percent of the direct-to-consumer business in 2003 to 65 percent of it in 2014.

While other companies had been experiencing a downturn in sales and many closed up shop, Stark was not only surviving, it was thriving. Sales were up 30 percent compared to 2007. The revenue increase marked a return to prosperity after many rebuilding years.

They say wars and elections are never good for the nursery business, because no one is paying attention to the catalog. Conversely, a downturn in the economy can sometimes be good. Between the big-bank collapse in 2008, and high prices at the pump and in the produce aisle, home gardeners were fed up. They started growing their own dinner. Sales of vegetable seeds, tomato transplants and fruit trees soared.

The website quickly became a major revenue source; it is difficult to imagine that only two years before, it was virtually non-existent. Kyle also successfully accomplished a re-branding of Stark Bro's with the help of the Meredith Corporation in Des Moines, Iowa, including the current logo.

The 500-year flood destroyed a nearby church and thousands of Stark trees in 2008.

Just when everything was going so well, there was a flood of the 500-year kind. It devastated the fields, and killed tens of thousands of trees. Five years later, there would be another flood almost as bad.

From 2001 through 2010, the company's progress wasn't very visible to the outside world, but the engine was steaming. So much had been lost between the sale and F&G bankruptcy, and to say that there was considerable ground to make up was a gross understatement. Maintenance had been neglected, the warehouses needed repair and most of the infrastructure had to be rebuilt. The time taken to fortify the foundation was well worth the reward, which would come along in a couple of short years.

Employees met in September of 2011 to celebrate 10 years under the new ownership. Elmer Kidd, Chief Production Officer, generously put on an old-fashioned fish fry. Tim and Cameron graciously honored "the budders", the five workers who volunteered to do field maintenance work prior to budding, and the six who volunteered to bud the trees in the summer of 2001. Without these dedicated people, the company would have died a certain death.

A new room opened that year at the Louisiana Area Historical Museum to showcase the history of Stark Bro's and the family. The exhibit includes machines used in the operation, old signs, memorabilia and the couch that Governor Lloyd Crow Stark once used in Jefferson City. Many of Luther Burbank's papers are also there. The room really symbolizes the company's endurance and invincibility.

Kyle David Brown

An Illinois truck driver got on the road early on the day of December 14, 2011. He was hauling a log skidder on a long trailer, traveling west on Highway 54 to go to the Missouri side. To get there, he had to cross the infamous 86-year-old Champ Clark Bridge, with its two narrow lanes and rickety beams.

According to the Illinois State Patrol, at 8:15 a.m., at the exact moment Kyle Brown's 2011 BMW SUV was crossing the bridge in an eastward direction, the chains holding the skidder snapped and the machine slammed onto the eastbound lane. It was too late to brake, and Kyle's SUV collided head-on with the machinery.

"We heard the sirens," remembers Barb Brewer. "People were talking that there had been an accident on the bridge. They were saying it was a black BMW, and I thought, 'It's Kyle.' So we were called into a meeting and told what happened. It was just so horrible."

His passing devastated the employees and threw the comfortable, routine hum of the business into chaos. Bobby Birks, the IT manager then, remembers it well.

"Kyle still had responsibilities [in IT] but he also handled network admin, budgeting and purchasing, so I inherited a lot of those responsibilities. He left us with a lot of momentum to move forward."

"Everyone on the team really stepped up their game to keep things running smoothly," said Jacob DuBray, Stark's website developer. "Whether it was Dennis pitching in with forecasting and promotions, Will with the catalog, or Debbie Epley. There was no way Bobby and I could have done everything alone, and thankfully we didn't have to. I really can't give enough credit to everyone here for helping get us through those very tough times."

Kyle, who was 40, died at the scene. He left his wife, Kristen Brown, and their two children. The truck driver was cited with improper lane usage, and since then, the speed limit has been reduced, as has the weight limit. The Champ Clark Bridge was closed for eight days for repairs.

Kristen sued the driver for wrongful death, claiming the truck was exceeding the speed limit and that the driver failed to keep a proper lookout. She was awarded $3 million by a federal jury in 2014.

Kyle's legacy will clearly be the technology vision he overlaid on the strategic plan. He had a fervent belief in the power of it, and knew the potential it held. Since that time, Stark has become one of the most technologically-advanced companies in the industry, and the rewards for following Kyle's vision have paid off many times over.

Ken Lane,
Chief Marketing Officer

Ken Lane, a notable name in the direct marketing field, was recommended as someone who might be able to help Cameron and Tim find someone to replace Kyle, not an easy task. Cameron had run into Ken at various industry conferences, so they were acquainted. To the surprise of both owners, Ken asked if he could toss his hat into the ring, and he was eventually hired.

Said co-owner Tim Abair: "The addition of Ken has been the most important factor in transforming Stark into a company that is on very solid financial footing. He brings industry knowledge, forward thinking and challenges Stark's approach in such a productive way."

Ken had a wealth of experience as a long-time strategic direct-marketing consultant, solving problems for large national clients. He believes that planning, forecasting and segmenting based on data is smarter and cheaper, and it is that philosophy forms the basis of his management springboard.

"I've been preparing for this role my entire career," he said. "It's akin to chess pieces being moved over time — some by me, some by others, some by God. I don't take any of this for granted and I will embrace this day, this opportunity, these people and this place. Why has Stark had these past few great years? I tell my team: 'look to your left, look to your right. That's who made it happen.'"

It is also Ken's purview to make sure that the technology stays on the right track. Kyle had already made the company one of the more technologically-enabled entities in the entire industry. Now, the company's systems are intricate, flexible and even intuitive, and Stark generates more than two-thirds of its business through its website because of it.

Ken said: "'We've always done it that way' has been replaced with 'How can we…?,'a seismic shift in approach from the old days."

Like many direct-to-consumer marketers with a catalog in its history, Stark Bro's had to evolve from print-based marketing to "all things E (for electronic)." Since 2010, Stark Bro's has grown to volume levels not seen since the 1970s, by mixing promotional e-mails, pay-per-click advertising, search engine optimization and social media. Gone are the days when they printed tens of millions of catalogs every year; gone are the days when 50 percent of the spring business was generated by February 15. No longer are mailed-in orders almost half the business.

The Golden Delicious stamp (2013)

At last, in 2013, the company was profitable again. The sigh of relief could be heard in Illinois.

That year would be full of surprises. First, the state fruit of West Virginia, the Golden Delicious apple, became a 33-cent stamp. The apple also celebrated its 100th birthday.

Then, Stark Bro's acquired J. E. Miller Nurseries, of Canandaigua, New York, its east-coast competition. With Miller's resources and 75 years of experience, Stark Bro's became the world's largest online direct-to-consumer nursery.

"When my brother David passed away unexpectedly, I knew it was time to retire," said J. E. Miller Nurseries CEO John Miller. "One of the first things I did was call my longtime friend, Cameron Brown. The quality of their products and the care they take with their own customers made it an easy decision. I know David would be pleased with this arrangement."

Cameron Brown (left) and John Miller

The buyout of Miller's "cold-hardy" stock has rounded out Stark's inventory and significantly added to revenues by opening up the Northeast market.

The following January, Stark was offering 40 of Miller Nurseries' "Canandaigua-Quality" trees along with selected varieties of berries and grapes. Customers are happy.

300 solar panels were installed in 2014

In yet another technological advance, Cameron and Tim initiated a project to "go greener" and install 300 solar panels on the administrative, greenhouse and warehouse buildings. The decision made Stark Bro's the only major online nursery to implement solar power as an integral energy source for its business operations.

Three arrays of 100 panels produce an average of 96,000 kilowatt hours of power annually. According to the U. S. Energy Information Administration, this is the equivalent of saving the electricity used by nine homes per year.

In June, Tim Abair presented the Milwaukee-based Growing Power, Inc. with a huge donation of 3,000 trees to help urban communities grow their own sustainable food sources. Instead of heading for the burn pile, the trees were used by the organization to turn empty urban lots into organic mini-orchards, providing good food for the local community, helping the environment and providing jobs for people to tend the trees.

Will Allen of Growing Power Inc., gets ready to plant one of 3,000 trees donated by Stark Bro's in 2014.

Stark Bro's is vastly different today than it was even 10 years ago. It is leaner and more efficient. Fewer people can be more productive because of technology advances; through better tracking of sales and forecasting, there are far fewer unsold fruit trees.

A small fortune has been invested in maintenance upgrades: new warehouse cooler compressors, insulation, updated electrical wiring and greenhouse boilers. New packaging methods, like Elmer's 4x10 pots, mean stronger plants and fewer replacements. The company has grown 100 percent, and most importantly, has started turning a consistent profit again. There is a need for a new warehouse, a very good sign.

And even today, while technology may be bolting forward in the office, in the fields, it is still the 19th century. Many of the heavy field chores have been mechanized, but all of the budding, pruning and tending must still be done by hand. Some people like it that way.

As successful as the Starks were, it can be comfortable to stay with old business practices, which they did to a large extent. Opportunities for forward movement and efficiencies were overlooked, but that was remedied when Tim and Cameron, both from a technology background, stepped in. The owners also gave employees more control over tasks, and along with it, more accountability.

It would be impossible to count the number of trees that have been sold in the last 200 years. Back in 1966, Lloyd Stark Jr. estimated that the trees sold in the last 75 years, if planted at the usual 20 feet apart, would circle the globe more than ten times.

Ken Lane, Chief Marketing Officer, believes the company has stayed alive because it has embraced change and technology. "We get stuff done. We work well together, respect one another and are not afraid to own our mistakes," he said. And they have one credo, cliché that it is: "Take care of the customer and the customer will take care of you.

"It is worth stating that the customer's role in our success is tantamount, as well. They know that there is something about buying fruit trees and berry plants from the grower, not a couple guys in an office park who buy someone else's trees, mark them up tenfold, then try to sell them at 50 percent off, which is still often a lot higher than Stark Bro's. Today's consumer is a lot smarter than that."

The Field Production division still carries on the 200-year-old traditions of stringent research and variety evaluation, production and highly personal service to their customers. Some things will never change. New grafting methods, new rootstocks, new products (like a hardy Asian pear and a pecan that's the shape of a walnut) are always in the incubator.

Marketing, by its very nature, is constantly changing. Stark still wants to reach the "millenials," i.e., the younger generations, as Clay tried to do many years ago. Social media is the avenue by which to do that. Incorporating it into Stark Bro's marketing strategy was originally Kyle Brown's idea. He believed that interacting with customers on a personal level was the way of the future. Kyle's father's goal is to have younger generations "trade a flat screen for a flat spade." Communication between people is now 140 characters on Twitter and status updates on Facebook; the key is that it's two-way communication. Stark Bro's plans to use it.

Bobby Birks, now Senior Software Engineer, says his area is now working on upgrading URSA to a new framework called PANDA. "We're a lot closer to technology that exposes the web to us, rather than hides it from us. That's really the core of the Panda project," he said.

The Next 200 Years

Between Lou, Cameron and Kyle Brown, Stark Bro's is once again a family-run business. As history illustrates, almost all nurseries are family-owned. But this is an era in which children are less likely to walk in their elders' footsteps, less willing to make the hard sacrifices. The question is, can another dynasty take the company through another 200 years? There is a special interest from another of Cameron's sons, Randy, in joining the business. Time will tell.

The number of mail-order fruit-tree companies has dwindled from hundreds in the early 20th century to about five. Even Jackson & Perkins (the famous rose company) has gone bankrupt twice. Stark Bro's is likely to survive because they grow their own trees, they grow their own rootstock and they are the go-to experts in grafting and the science of hybridization. And don't forget the employees.

Add to that the fact that people are becoming more educated about the food they eat, and are much more interested in knowing where their food comes from. Those trends, paired with Stark's talent for innovation and marketing, will keep them at the forefront.

"As the company looks forward, we recognize that the direct-to-consumer model will plateau," says Ken Lane. "When that time comes, we expect to have developed a Stark-branded line of trees and berries for the upscale, independent garden-center market. You won't see us trusting Stark products to be properly handled at any big-box stores; that was a lesson learned long ago. I think we are a decade away from that, but we are planning for it nonetheless."

Eventually URSA will go away. The next tech target is planning systems. Right now, IT is melding the inventory tracking systems between the fields, the warehouses and the greenhouses, so one hand knows what the other is doing. The goal is to have "systems and processes so good that they spoil people." Expanded mobile-marketing efforts and on-the-go payment, along with wider use of iPads in the field and greenhouses are among the many items on the future-improvements calendar.

Ecological efficiencies like internal recycling and rainwater reclamation for the greenhouses will save money and lessen their environmental footprint. Organic trees are on the horizon.

Ridding orchards of spraying will be a tough proposition, however. Customers often ask: "How can I get good apples without spraying?" The industry punchline is, "Cut down the tree and buy apples from the local fruit stand." Perhaps that will evolve and organic treatments will solve all pest and disease problems.

American consumers' tastes will change. Now, they want Galas, Fujis, Goldens and Granny Smiths. Tomorrow, SweeTango and Jazz. Twenty years from now, no one knows. International buyers will continue to provide demand for Red Delicious that has diminished in the U. S. Someday the famous red apple can be bred back to its truly delicious original self.

Some direct-marketers feel that the printed catalog will go away. Ken Lane is not naïve enough to believe that might be the case someday, but thinks that will happen only if the USPS makes mailing prohibitively costly. The entire concept of printing and mailing requires a solvent USPS, or some other financially-feasible means of delivering paper marketing materials.

The reputation and heritage of Stark Bro's will be preserved above all, if Cameron Brown has anything to say about it. The owners view their role not as traditional corporate investors, but as stewards of the Stark name. "I feel like I'm guarding the Declaration of Independence," Brown said.

If belief in the seeming immortality of this company could be summed up in a few words, it would be these, from Ken Lane:

> *"Stark Bro's has been a slice of Americana for the last 200 years, whether you knew it or not. And whether you believe it or not, it will be for the next 200 years."*

While there will likely be no Stark family members employed at the company, their presence will still be felt daily, as will Luther Burbank's. The fervent loyalty to the company's heritage remains; it is apparent from the CEO to the field workers that they exist not only to work, but to be good protectors of the legacy and remarkable history of the company.

Quoting famous entrepreneur Robert Kiyosaki, "The size of success is measured by the strength of desire, the size of your dream, and how you handle disappointment along the way."

Terry Stark (not a *Stark* Stark) provided that quote. Terry is the self-described "tomboy right off the farm" who so gratefully credited the company for her personal growth. She embodies Kiyosaki's quote, and is a living, breathing example of why Stark Bro's Nurseries & Orchards Co. is still here, after 200 years.

Stark Bro's Awards and Achievements

Year	Awarded by	Award	To
1920	Indiana Horticultural Society	1st prize	Golden Delicious
1919	American Pomological Society	Wilder Medal	Golden Delicious
1926	American Pomological Society	Wilder Medal	Starking® Delicious
1972	American Pomological Society	Wilder Medal	Paul Sr.
1960	American Pomological Society	Paul H. Shepard	Paul Jr.
1990	Missouri Centennial Corp.	Honor Roll	Stark Bro's 100th
1992	MO Chamber of Commerce	Agribusiness Pacesetter	Clay Logan
2003	MGA	Green Thumb	RPM™*
2005	MGA	Green Thumb	Prime-Jan®/Jim®
2009	Direct Gardening Association	Green Thumb	Golden Delicious Collection for Children
2012	Governor Jay Nixon EZ Start® Bottomless Pots	Governor's Award for Agricultural Achievement	Elmer Kidd
2014	Louisiana Country Colorfest	Best of Show	
2014	Winter Wonderland	Best of Show	

*Root Production Method

Elmer Kidd, receiving the Governor's Award for Agricultural Achievement in 2012

Stark Bro's in the Community

Over the years, Stark Bro's and their employees have contributed thousands — in both hours and dollars — to the local hospital, YMCA, area school, the Rotary Club and local chamber events. Below are just a few examples:

In 1993, Missouri saw some of the worst flooding in its history, with 30,000 square miles affected for as long as 174 days. The following spring, Stark donated more than 5,000 trees to surrounding towns to help them get back on their feet.

Clay Logan assists Kids in Motion participants in planting trees provided by Stark Bro's Nursery at the Riverview Cemetery. (Summer, 2008)

The Employee & Community Relations Committee raises funds to help support local charities through events such a raffles at our Service Award dinners, 50/50s or perhaps a meal that is made by and sold to the employees. Raffle prizes are mostly donated by the Company using AMEX rewards points. The charities we contribute these funds to are The Backpack Program, Food Pantry, and the "Spirit of Christmas" (Adopt-a-Family) program.

The Louisiana Press-Journal and Heart-to-Heart Community Outreach have long-sponsored a "The Spirit of Christmas" program to collect gifts, food and cash donations for the needy during the holidays. Stark Bro's employees have adopted several families over the years.

Many Stark Bro's employees participated in a "Paint The Pumpkins Pink" fundraising event to raise breast-cancer awareness. The proceeds were presented to Pike County Memorial Hospital for their free mammogram program in 2014.

Growing Power, a dynamic urban farming organization headed by CEO Will Allen (named one of TIME Magazine's 100 Most Influential People In The World), received a donation of more than 3,000 Stark Bro's fruit trees in the summer of 2014.

The trees, which were personally presented by Stark Bro's co-owner Tim Abair, are to be planted on over 2,500 lots in and around Milwaukee and Madison, Wisconsin, providing a safe, sustainable food source for area residents.

The Stark Bro's Employee Relations Committee held a fund-raiser for the Buffalo Township United Fund at the Stark Bro's Garden Center.

In 2011, Stark was a main sponsor of the 10,000 Garden Project, organized by the Missouri Department of Agriculture. The effort was to get 10,000 Missourians to grow a garden full of fresh, wholesome food, with prizes awarded at each 1,000-garden milestone.

Tales from the Orchards

The most important thing happening at Stark Bro's can't be seen in office buildings, ledgers, letters from happy customers or financial statements. It's the people who work there.

Over nearly 200 years of existence, the Stark Bro's Nurseries & Orchards Co. put down long roots in this small town by the Mississippi. When it fell on hard times, it found that its employee's loyalty ran just as deep.

As many as five generations of some families have worked for this nursery; they have raised families together, mourned deaths together and worked side-by-side, in some cases for decades. There are 19 employees who have worked there for 20 years or more, who still want nothing more than to wake up and come to work at Stark Bro's every day. Here are a few memorable quotes from the "old-timers."

Bob Patrick

Bob Patrick: "When I think of the nursery I think of extremes: 117 degrees when we were budding; 5 below trying to get cuttings in the winter. OSHA doesn't allow us to do that anymore."

Jack Alexander former VP and General Manager: "These people come from nothing, but will give you all they have. Honorable people doing honorable work, caring for their families and fellow workers. Providing plants that brought great pleasure to our customers. I absolutely loved working with them."

Jack Alexander

Terry Stark

Terry Stark (no relation): "In the old days there was a big metal wagon that would come around with water, and you'd get a drink with a dipper. Jim Kanaley chewed [tobacco] and we always tried to get to the dipper before Jim."

Barb Brewer: "My son left work here one night, and was killed in a car accident. Some of us bore our families here, and some of us bury our families. We're all Starks, even if none of us are actually Starks, if you know what I mean. For some of us, this is our family. We come in here and cry with our friends. The support I got when I lost my son was unbelievable."

Barb Brewer

Ginny Andrews

Ginny Andrews: "There used to be a cafeteria upstairs in what is now the Garden Center. Our meals were 35 cents. Miss Elliot would always cook the same delicious combinations: fried chicken, a baked potato and slaw. When she had salmon croquettes, broccoli and cheese, and hot rolls, it was just delicious. People would come in from the fields, all cold and hungry, and eat until there wasn't anything left."

Cameron Brown, President and CEO: "I believe Ginny Andrews has been at Stark since Adam and Eve. She looks 50 but must be 80. I think she runs on nuclear power."

Cameron Brown

Elmer Kidd

Elmer Kidd: "One day when I was with Paul Jr. I mentioned Carl Ince and Edgar Keith, who was a nurseryman. Paul Jr. said, 'They're both Homer-Reed-trained nurserymen. They're both Homer Reed disciples.' I remember that so well."

Marian Kuntz: "There is an ongoing debate as to whether or not a time capsule was buried at the 175th anniversary (1991) on the grounds near a tree and a marker. Some think we did; some swear we didn't. All I know is that there was definitely a tree and a marker (both of which are long gone)."

Marian Kuntz

Vera Murr

Vera Murr: "When the bankruptcy happened, there was a deluge of phone calls. Some asked about their order. Others wanted to say they were sorry. I had one gentleman call me twice a month to feed his hope that Stark Bro's would reopen."

Walter C. Logan II

Walter Logan II: "My dad was on the TV show "To Tell The Truth" in 1966. Bud Collyer was the host. I was a freshman in high school and wanted to watch it, so did the principal, so did everybody else, so they brought a TV into the cafeteria and the whole school watched it. I got a front row seat. It was then I realized that my family was pretty well-known."

Terry Stark: "Now we ship in cardboard boxes, but we used to ship large orders in elephant hide and brown kraft paper. You would really have to manhandle them to get them wrapped."

Barb Jenkins

Barb Jenkins: "I remember when I was a teenager and worked at Bankhead's (a locally famous candy store and restaurant) in Bowling Green, Missouri. Miss Bankhead told me, she said, 'if you get up three days in a row and don't want to go work, then that's a good sign that's not the job for you. You're not helping yourself, you're not helping the company and you're not helping your customers.' She also said 'you're always going to have a day now and then when you don't want to go,' but I've never had that. I love my job so much."

Lita Eatock: "I had just been offered a job at the nursery. I went back to the marina where I was living upstairs over the office/store area. As I walked into the store, I saw a 'scuzzy-looking' guy rummaging through a box of boat shoes. I said to the storekeeper, 'You better watch that guy. He might rip you off.' The storekeeper, who knew the person, said to me, sort of chuckling: 'That is the president of the company that just hired you!' John liked to 'dress down' on his off-days, especially when working on his houseboat."

Rose Marie Buchholz

Rose Marie Buchholz: "In 200 years from now, I will be one of the little stars up in the sky, looking down at the fields and the people at Stark Bro's They will still be here."

Barb Brewer: "We were counting grapes in the greenhouse. There were so many trees and they were all mixed up; it would have been easy to just count them as one. But we can't do that. So we sorted the varieties and got it done. It's just not 'us' to take the easy shortcut. It's not what we do."

Kenny Harrison

Kenny Harrison: "My best memory was being up on the hill with the old-timers. It was cold, and we built a big old bonfire to stay warm. We'd try to keep our feet warm and start burning our boots without even knowing it."

Marian Kuntz: "After John Logan retired, he was still Chairman and would come out to the office on Thursdays after Rotary. He would park in the Visitors area. One day the Customer Service Manager, Sandra Reed, backed into his car. She jokingly convinced him it was his fault, because he shouldn't have parked in the Visitors area. John was such a quiet person, who spoke as few words as necessary. He never held her responsible. He was such a kind man."

J. P. Wooten with Clay Logan and PIII

J. P. Wooten: At 107 years old, J. P. was the oldest former employee. He knew four generations of Starks.

Walter Logan II: "The catalog was being circulated among a group of six or so for proofing, and came back to me for approval. Unfortunately there were still a few typos: 'Radiant Flowering Crab 'was 'Radiant Flowering Crap'. "Bear Facts"'was 'Bear Farts' and 'Exotic Hardy Kiwi' was 'Erotic Hardy Kiwi.' How many sets of eyes were on that catalog? Makes you wonder how intentional it was. Still makes me laugh."

Loyalty Above and Beyond

Here's to the "Old-Timer" employees who served a total of 30 years or more. (Some are still employed.) Personnel records of older employees, some who worked at Stark Bro's their entire lives, were lost in the sale to F&G, but a concerted effort was made by current "old-timers" to recall all of their names.

Virginia Andrews
Albert Bass
Elsworth Bass
Preston Beck
Ricky Blackstun
Mary Frances Boyd
Ralph Brawner
Delphine Brawner
Barbara Brewer
Margaret Brown
Odetta "Ma" Brown
Rose Marie Buchholz
Oralee Calhoun
Kathleen Calhoun
Hurley Carr
Jerry Clarkson
Birdella Claus
Juanita Clifton
Gordon (Gilbert) Cohea
Elba Colbert
Mable (Angie) Colbert
Florence Crouch
Lydia Daniels
Artie Dill
Ada Dorwood
Bill Duckworth
Bill Eddington
Omer Eddington
Claudean Eddington Goodman
Ernie Elder
Wes Elliott
Annie Elliott
Deborah Epley
Herschel Estes
Norma Jean Evick Stone
Martin Fisher
Portia Forgy Richter
Opal Francis
Arthur Francis
Gussie Fritz
Loretta Fuhrman
Emily (Laveta) Garner
Everett Geery
Dean Gooch
Ida Green
Harry Guengerich
Nellie Hagen
David Hall
Dola Hammett
Ed Harpole, Jr.
Kenneth Harrison
Richard Harrison
Sue Harrison
Geraldine Hartsock
Merle Hartsock
Lawrence Henderson
Hobo Hickman
Carolyn Hobbs
Glenn Holman
Venda Houchines
Mildred Ince
Carl Ince
Mildred Ince
Martha Ingram
Darlene James
Barbara Jenkins
Wendell Jennings
Margaret Jennings
James Kanaley
Edgar Keith
Kenneth Kennedy
J.W. Kennedy
Tom Kennedy
Elmer Kidd
Patricia (Ann) Kindle
Ray Kindle, Jr.
Ray Kindle, Sr.
Lloyd Kuna
Nona Kuna
Marian Kuntz
Don Lewis
John Logan
Floyd Maples
Bob Masterson
Pearlie McChristian
Orval McChristian
Charles McChristian
Julia McCune
Sam McDonald
Raymond Miller
Ruth Miller
Nina Mitchell
James Mobley
Rondal Moore
Lola Moore
Bob Mosley
Raymond Mosley
Paul F Motley
Vera Murr
Ada V. Murr
Jerome Myers
Norman (Jim) Myers
Pauline Myers
August (Gus) Owens
Richard Owens
Ella Patrick
Robert Patrick
Hallie Peabody
Otto (Glen) Pence
Juanita Potter
Leonard Pounds
Ira Reed
Homer Reed
Paul Reller
Betty Richmond
Jo Riley
Kathleen Riley
Mary Rose
Lorene Salling
Errett (Bud) Sapp
Dora Sapp
Virgil Sapp
Cloyde Saxbury
Mrs. Cloyde Saxbury
Mary Lois Schneidler
Calvin Short
Kelly Short
J. B. Shy
Rose Silvey
Frank Smith
Ethel Smith
W.T. Smith
Gene Smith
Margaret Spears
Terry Stark
Paul Stark Jr.
Cheryl Sutton
Glenn Thomas
Nora Thurmond
Clarence (CG) Wahl
Bob Wahl
Maude Wilson
Verne Windmiller

Acknowledgments

Governor Jay Nixon
Rebecca Baker
Dr. Susan K. Brown
Russ Pinto
Erin Sheffield
Emily Jaycox
Jacob DuBray
Carl Ince Family
Jack Alexander
Dennis Moronakis
The Harrisons
Barb Jenkins
John Stoeckley
Frank Absher
Vera Murr
Bobby Birks
Rose Marie Buchholz
Tabitha Rardon
Jane S. Smith
Will Carson
Michael V. Geary
Elmer Kidd
The Kindles
Sarah DePass
Bob Patrick
Barb Brewer
Joe Weston, Jr.
Dennis Petretti
Luther Burbank Home & Gardens
Martha Sue Smith
Carol Burbank
Mary Rochelle
Maureen Tomczak
Carol Baker
Rep. Jim Hansen-Missouri
David Johnson
Julie S. Higginbotham
Ginny Andrews
Ricky Blackstun
Marilyn (The LOC librarian)
Caren Diane Stark Brown
Jim Law

First and always, I thank God for the opportunity to write a book about these extraordinary people.

Most heartfelt thanks to Clay Stark Logan, Walter C. Logan II and their wonderful mother, Mary Alice, for their invaluable donations of time, historical knowledge, family photographs and memories. This book simply could not have been written without their thoughtful contributions — or their intrepid ancestors.

Immense thanks to owners Cameron Brown and Tim Abair, for the unfettered access to company archives, employees, facilities and themselves.

My personal gratitude is owed to Ken Lane, Stark Bro's Chief Marketing Officer, who has been a mentor, colleague and friend for more than 20 years. He has profoundly influenced my life and the way I think, and I am a better person for it. I hope he donates his brain to science; the doctors will be astounded.

Especially enthusiastic applause for Stark employees Debbie Epley and Marian Kuntz, who spent hours researching, digging and dusting off photos and documents from the Stark Bro's archives, not to mention wracking their memories to help solve long-standing mysteries. You were indispensable to this project.

In 1966, a booklet was written for the 150th anniversary entitled, "The Stark Story," by Dickson Terry, in conjunction with the Missouri Historical Society. On that occasion, former Stark Bro's Chairman and Missouri Governor Lloyd Crow Stark remarked:

"There have been so many splendid men and women not related to the Starks who have played leading parts in the development and success of the Stark Nursery over these many years, [and the length of this book] prevented us from relating many incidents and constructive accomplishments of these loyal and dedicated members of the Stark Bro's organization. It is hoped that a full account will someday be written so that the story of these loyal leaders, who have played such vital parts on the team, can be adequately told."

While this book is 200 pages long, there was still not room to include the significant contributions of these Stark Bro's employees, and so they are acknowledged here:

John Ganzer • Dola Hammett • Darlene James • Bob Wahl

Well-Wishes From Dignitaries Far and Wide

Missouri Congressman and former Speaker of the House James "Champ" Clark (1850 — 1921)

From a speech to the House of Representatives in 1906:

"Most of the early settlers of Missouri were soldiers or sons of soldiers. Among them was Judge James Stark of Bourbon County, Kentucky, who served under General Richard M. Johnson at the Battle of the River Thames. Judge Stark came in the prime of his manhood from Bourbon, one of the finest counties in Kentucky, to Pike, one of the richest in Missouri, reared a family which is still numerous and powerful, and laid the foundation for the largest nursery in the world, now conducted by third and fourth generations of his descendants. We sometimes build more wisely than we know. He little dreamed while riding horseback from the far-famed Kentucky bluegrass region, carrying in his old-fashioned saddlebags the scions with which to start a nursery, and from which grew the first grafted fruit trees ever propagated on the sunset side of the Mississippi, the name Stark honorably familiar not only throughout America, but beyond; and that 90 years after he planted his trees, his grandchildren and great-grandchildren would be shipping trees beyond the seas; but such is the truth of history. It is really a pity that he cannot return to earth long enough to witness the magnitude of the business he started. Verily, his works do follow him."

Missouri Governor Jay Nixon:

"One of the many pleasures and privileges that Georganne and I have enjoyed as Governor and First Lady is meeting home-grown, family-owned businesses like the Stark Bro's nursery.

When James H. Stark moved west and started his business here in 1816, no one could have imagined the influence the business and family would have had on our state and nation. As the industry leader in fruit-breeding and tree production, the products they have introduced to the world are now mainstays in our yards, kitchens and grocery stores.

The Stark family has made its mark on Missouri in many ways, including Lloyd Crow Stark who served as the 39th Governor of Missouri from 1937 to 1941. We have continued to carry on that tradition by recognizing Stark Bro's Nurseries & Orchards in 2012 with the Governor's Award for Agricultural Achievement, which honors outstanding farmers, growers and processors for their work with agricultural commodities and their communities.

This past year, Georganne and I were honored to join Elmer Kidd at the Missouri's Governor's Mansion where we planted six different types of Stark Bro's fruit trees, including Crimson Crisp, Red Delicious, and Golden Delicious apple trees and blackberries so that Missourians can see and taste a bit of Missouri history.

As Stark Bro's celebrates their 200th anniversary as a family business here in Missouri, we want to congratulate them on their success and thank them for being part of Missouri's proud agricultural legacy."

Missouri State Representative Jim Hansen, District 40:

"As the country's oldest nursery, they have touched the lives of so many people in the industry. Stark Bro's is truly a living artifact of American horticulture. I want to wish them the highest congratulations on their many accomplishments and long-standing history of success, which includes the best Red and Golden Delicious apples on God's green earth. Stark Bro's provides us with validation that the 'American Dream' is still alive and well!"

Bart Niedner, Mayor of Louisiana, Missouri:

"In many ways Stark Bro's is the bedrock-foundation upon which Louisiana stands. As Mayor, it is impossible to meaningfully plan for the City of Louisiana without also focusing on Stark Bro's Nurseries. Our histories are dependently entwined; our current vitality is mutually reliant and our future will best be served by our continued synergy. It is more than a simple economic reality: our history, self-image and sense of contribution to the larger world extend from a core relationship between this stalwart business and our historic community. As we each approach our bicentennial, it is with great mutual strength that the City of Louisiana conceives its future success and prosperity. The city could have no better ally than the one we find in Stark Bro's."

Carol Burbank (Luther Burbank descendant)

"Luther believed the Starks were the individuals most appropriate to store and continue his groundbreaking work. Thanks in part to the Stark Bro's, the work my ancestor will continue to nourish and bring joy to generations of people.

Dr. Susan K. Brown, Ph.D.
Associate Dean CALS and Director NYSAES
Professor, Cornell University

"Stark Bro's Nurseries & Orchards Co. has had a strong legacy since its 1816 establishment. It is amazing that one nursery could have had the foresight to recognize the value of what would become two standards, Stark Red Delicious in 1893 and Stark Golden Delicious in 1914.

Stark Bro's passion for plants and their propagation gained the notice of an early plant breeder. The inimitable Luther Burbank developed over 800 varieties and selections. Luther Burbank's plums were vital to the establishment of the plum industry in California. When Luther Burbank was nearing the end of his career, he entrusted Stark Bro's with most of his remaining plant materials. When a man as passionate about plants as Luther Burbank transfers that trust to another entity, there is no higher compliment. Stark Bro's was viewed as deserving of that trust.

Tenacity, dedication and especially their passion for plants has allowed Stark Bro's Nurseries to weather many challenges, and remain poised to continue their strong legacy of providing new and proven varieties to growers and home gardeners. Congratulations on your 200th anniversary!"

George Ball, Chairman and CEO, W. Atlee Burpee Company

"Stark Bro's and Burpee have each worked tirelessly toward the innovation and betterment of the world through horticulture. It was David Burpee who carried the mantle, upon his cousin Luther Burbank's passing, of furthering Burbank's unfinished experimental work, the rights to which were acquired from Stark Bro's in 1931. Just as Stark Bro's applied their passion toward propagating Burbank's fruits, David Burpee worked to bring many varieties of new flowers to American home-gardeners. Burpee and Stark Bro's complementary paths have been paved with a dedication for excellence. We congratulate Stark Bro's on the milestone of their 200th anniversary."

Bibliography

"Advertisement for Wm. P. Stark Nurseries." *The Country Gentleman*. January 27, 1917.

"All About Apples." U.S. Apple Association. Accessed February 5, 2015. http://www.usapple.org/index.php?option=com_content&view=article&id=19&Itemid=17.

American Seedsman. December 15, 1921.

Schenectady Gazette. "Apple trees "balk frost by ice coat." April 27, 1938, 14.

Bauer, Grace S. *Reminiscences of Washington Stark of his father, Judge James Hart Stark*. 1911.

Brown, D. "An apple from the world's largest nursery." *Globe Democrat Sunday Magazine* (St. Louis), October 9, 1966.

Brown, M. "Keeping the nursery local." *Pike Press* (Pittsfield, IL), October 28, 2001.

Burbank, Luther, et al. *Luther Burbank: His Methods and Discoveries and Their Practical Application*. New York [N.Y.]: Luther Burbank Press, 1914.

Buschmann, K. "Missouri Business – October 2012 : 100 Years Of Missouri Business." Last modified October 2012. http://www.mlppubsonline.com/article/100_Years_Of_Missouri_Business/1201773/129354/article.html.

Carey, Christopher. "Stark Bro's Tries To Grow New Business After Revival." *The St. Louis Post-Dispatch*, November 9, 2001.

Collins, J. R. "Notes and Queries - J. Edward Maybee (1866-1941." 1866.

Cristinzio, M. "Wholesale Nursery Design." Lecture, December 9, 2008.

Crow, J. F. "Plant breeding giants: Burbank, the artist; Vavilow, the scientist." *Genetics*, no. 158 (2001): 1391-1395.

Daily Express, The (San Antonio, Tx). April 6, 1906.

Daily Journal-Gazette, The (Mattoon, IL). May 26, 1914.

Dean, B. "End of the Road for Foster Gallagher - Direct Marketing News." Last modified July 24, 2001. http://www.dmnews.com/end-of-the-road-for-foster-gallagher/article/73884/.

Del Franco, M. "The Top 20 Stories of the Past 20 Years | Multichannel Merchant." Last modified April 15, 2003. http://bit.ly/1OllOzG.

"Documents | Tree Top Food Ingredients." Tree Top Food Ingredients | Fruit Suppliers. Accessed January 10, 2015. http://foodingredients.treetop.com/document.

Dreyer, P. *A gardener touched with genius: The life of Luther Burbank*, 2nd ed. Santa Rosa, CA: Luther Burbank Home and Gardens, 1993.

Engel, B. "Stark Brothers completes merger with Indiana firm - News - Hannibal Courier - Post - Hannibal, MO." http://www.hannibal.net/article/20100127/News/301279858#ixzz3BQt8mnTf.

Engle, Brent. "Seeds of growth: historic nursery keeps ahead of the times." *Hannibal Courier-Post*, February 24, 2010, 1A, 14A.

"F&G Hires Interim Chief at $21K a Week." Multichannel Merchant. Accessed January 15, 2015. http://multichannelmerchant.com/news/fg-hires-interim-chief-at-21k-a-week-26072001/#_.

"Family Business - The World's Oldest Family Companies." Accessed March 3, 2015. http://www.griequity.com/resources/industryandissues/familybusiness/oldestinworld.html.

"Florists' Review - Google Books." Last modified May 11, 1916. http://bit.ly/1FFiXAd

Florist's Review 43, no. 1093-1105 (November 1918): 77.

Florist's Review 33, no. 845-858 (n.d.): 84.

"Foreign Agricultural Service Import Data." USDA. n.d. apps.fas.usda.gov/gats/default.aspx.

Fowler, C. "The Plant Patent Act of 1930: A sociological history of its creation." *Journal of the Patent and Trademark Office* 82, no. 621 (2000): 44.

Frecon, J. *HortScience*, February 1979.

"Gov. Lloyd Crow Stark - Historical and Biographical Notes." Missouri Sec. of State Jason Kander. Accessed September 30, 2015. http://on.mo.gov/1RkWmKw..

Grote, Jim. "Conflicting Generations: A New Theory of Family Business Rivalry." *Family Business Review* 16, no. 2 (2003): 113-124. doi:10.1111/j.1741-6248.2003.00113.x.

Harris, Mary Kathryn, and Mary Iva Jean Jorgensen. *James Stark of Stafford County, Virginia and His Descendants*. Fort Worth, Tex. (3525 Scranton Dr., Fort Worth 76118): M.K. Harris, 1985.

Hannibal Courier-Post. "Harry Guengerich retires after 60 years." October 28, 1976.

Hauser, W. "Forward Thinkers." *Western Fruit Grower*, November 1999.

Herzog, Milwaukee Wisconsin Sentinel Journal, K. "SweeTango at core of apple war." Last modified September 9, 2010. http://www.jsonline.com/news/wisconsin/102597659.html.

Higginbotham, J. S. "Stark's new reality." *American Nurseryman*, July 15, 1991.

"History of Wilder Award winners." Accessed March 11, 2015. http://bit.ly/1M2cfkr.

Howard, W. "Luther Burbank's plant contributions." *Chronica Botanic, University of California College of Agriculture, Agricultural Experiment Station*, no. 691 (1945).

Howard, W. "Luther Burbank, a victim of hero worship." *Chronica Botanica* 9, no. 5-6 (1945).

Inc.com. "The top 25 oldest companies." n.d. http://bit.ly/1L98GfU

Internet Archive: Digital Library of Free Books, Movies, Music & Wayback Machine. Accessed February 4, 2015. http://archive.org. 1998-2015, www.starkbros.com

Jackson, Sam. "Gardens of Luther Burbank Now Idle; Seek New Genius." *Gettysburg Times*, September 30, 8.

Jacobsen, R. "Why Your Supermarket Sells Only 5 Kinds of Apples | Mother Jones." Last modified March 2013. http://bit.ly/1imrsgS.

Jones, D. F. "The life and work of Luther Burbank." Lecture, Spragg Memorial Lectures, Plant Breeding, Michigan State College, East Lansing, MI, 1937.

Joy, J. "The "Garden of America" - Early Days of the Fruit Industry in Southern Illinois." *The Illinois Steward*, Winter 2003.

Kalinoski, Glenn J. "Date Set for Auction of F&G Horticultural Titles." *Direct Marketing News*, August 2, 2001.

Kallen, Stuart A. *The War at Home*. San Diego: Lucent Books, 2000.

Keith, C. "Centennial History of Pike County." Speech, July 4, 1876.

Kevles, D. J. "Protections, Privileges, and Patents: Intellectual Property in American Horticulture, the Late Nineteenth Century to 1930." *Proceedings of the American Philosophical Society* 152, no. 2 (2008): 207-17.

Kingsbury, Noël. *Hybrid: The History and Science of Plant Breeding*. Chicago: University of Chicago Press, 2009.

Louisiana Press Journal (Louisiana, MO). "Expansion to give local economy a boost." October 17, 2007.

Marra, J. "Golden Delicious apple discovered in Clay County." *The Herald Dispatch* (Huntingon, WV), September 1, 1969.

McEowen, B. "Stark Bro's blooms again." *Rural Missouri*, May 2004.

Milkovich, Fruit Growers News, M. "Litigants settle SweeTango dispute." Last modified November 4, 2011. http://fruitgrowersnews.com/index.php/magazine/article/litigants-settle-sweetango-dispute.

Missouri Highways. October/November 1971.

Missouri State Board of Agriculture. *The Missouri Yearbook of Agriculture: Annual Report* 4: 509.

Moberly Monitor-Index and Democrat (Moberly, MO). September 22, 2008.

"National Gardening Association :: Garden Market Research." National Gardening Association :: Garden Market Research. Accessed May 30, 2015. http://bit.ly/1hbkxxY.

National Nurseryman, The, January 1921, 68.

National Organic Program (U.S.). *Organic Food Standards and Labels The Facts*. [Washington, D.C.]: U.S. Dept. of Agriculture, Natural Organic Program, 2002. http://1.usa.gov/1KSfP0t.

National Register of Historic Places. *National Register of Historic Places*. [Washington, D.C.]: U.S. Dept. of the Interior, National Park Service, National Register of Historic Places, 2003. http://www.nationalregisterofhistoricplaces.com/. OMB NO. 1024-0018.

"Nursery Crops." Last modified 2007. http://bit.ly/1QLq76a.

"Ostertag Resigns As Foster & Gallagher CEO." Chiefmarketer. Last modified December 6, 2001. http://www.chiefmarketer.com/ostertag-resigns-as-foster-gallagher-ceo/.

Peck, B. "Missouri fruit trees." *The Backfence*, Summer 2011.

Pike County Genealogical Society. "Stark Cemetery." n.d. http://bit.ly/1JDidq5.

Reich, Associated Press, L. "Patents and trademarks: my plants broke the law." Last modified December 27, 2012. http://bit.ly/1MGVC0q.

Reich, Associated Press, L. "Some Red Delicious are more delicious than others | Deseret News." Last modified October 10, 2012. http://bit.ly/1O8pJB6.

Rodda, Kelli. "Social hours: Stark Bro's Nurseries & Orchards makes big marketing strides with social media." *NurseryManagementOnline.com*, September 2011, 17.

"Royal Gala Apples | Australian Apple Varieties | Aussie Apples." n.d. http://www.aussieapples.com.au/aussie-grown-varieties/royal-gala.aspx.

Scholten, H. J., H. T. Hartman, D. E. Kester, F. T. Davies, and R. L. Geneve. "Plant Propagation: Principles and Practices." 1997.

Schwadron, Karen. *Pike County, Missouri: People, Places & Pikers*. Marceline, Mo: Walsworth Pub. Co, 1981.

Simon, Ellen. "Struggling gardeners turn to yard-grown produce." *Telegraph Herald* (Dubuque, Iowa), May 25, 2008, 8.

Smith, Jane S. *The Garden of Invention: Luther Burbank and the Business of Breeding Plants*. New York: Penguin Press, 2009.

"Spotlight | National Inventors Hall of Fame." n.d. http://invent.org/inductee-detail/?IID=21.

Springer, Dan. "Foster & Gallagher files for bankruptcy." Hort.net. Last modified July 5, 2001. http://www.hort.net/lists/perennials/jul01/msg00017.html.

Louisiana Press Journal (Louisiana, MO). "Stark Bro's goes solar with 300 panels in place." July 2, 2014.

Stark Jr., Paul C. "The Interstem Tree: The Compact Tree of the Future?" *American Grower*, February 1972.

Summers, D. "Nursery Insight: Outlook for 2014." *American Nurseryman*, February 2014.

Terry, D. *The Stark Story: Stark Nurseries 150th Anniversary*. St. Louis, MO: Missouri Historical Society, 1966.

Tonge, P. "Back to grafting and budding basics in Hannibal, Mo." *Christian Science Monitor*, August 8, 1980.

United States. Agricultural Statistics Board. *Noncitrus Fruits and Nuts*. Washington, D.C.: [National Agricultural Statistics Service (NASS), Agricultural Statistics Board, U.S. Dept. of Agriculture, 1974. http://purl.access.gpo.gov/GPO/LPS1160.

United States. Department of Agriculture. Economic Research Service. Market and Trade Economics Division. United States. World Agricultural Outlook Board. *Fruit and Tree Nuts Situation and Outlook Yearbook*. Washington, DC: 2002. http://purl.access.gpo.gov/GPO/LPS54431.

United States. Department of Agriculture. Economic Research Service. *Organic Production*. [Washington, D.C.]: Economic Research Service, U.S. Dept. of Agriculture, 1900. http://purl.access.gpo.gov/GPO/LPS50382.

United States. Department of Agriculture. Office of Information. *1946 Victory Garden Program: USDA Materials and Activities*. 1946.

United States. Department of Agriculture. *Victory Garden Leader's Handbook*. 1943.

United States. National Agricultural Statistics Service. *2002 Census of Agriculture. Volume 1, Part 51, Volume 1, Part 51*. [Washington, D.C.]: U.S. Dept. of Agriculture, National Agricultural Statistics Service, 2004. http://purl.access.gpo.gov/GPO/LPS59160.

USDA: 2012 Census of Agriculture. United States. Dept. of Agriculture, 2013. http://www.agcensus.usda.gov/index.php.

"Verdict in Stark inquest." *The Florist's Review* 34, no. 859 (1914).

"Washington Apple Country." Gourmet Cider :: Home. Accessed February 12, 2015. http://www.appleorchardtours.com/hist06.htm.

Williams, H. S., ed. *Burbank, his life and works*. New York: Heart's International Library Co., 1915.

Wittenauer, C. "Branching out." *The Southeast Missourian*, January 19, 2003, 11B.

Woodward, K. *Quincy Herald-Whig* (Quincy, MO), November 6, 2001, 5A.

Yager, S. "The Awful Reign of the Red Delicious - The Atlantic." Last modified October 9, 2014. http://www.theatlantic.com/health/archive/2014/09/the-evil-reign-of-the-red-delicious/379892/.

Young, J. *Missouri curiosities: Quirky characters, roadside oddities & other offbeat stuff*. Guilford, CT: Globe Peguot Press, 2003.

CPSIA information can be obtained at www.ICGtesting.com
Printed in the USA
LVOW07*1625300116

471977LV00003B/8/P

9 780692 312919